FARMERS THAT HELPED SHAPE AMERICA

A Van Sickles Family History

Joseph Vincent Collins

Hamilton Books
A member of
The Rowman & Littlefield Publishing Group
Lanham · Boulder · New York · Toronto · Plymouth, UK

Hamilton Books
4501 Forbes Boulevard
Suite 200
Lanham, Maryland 20706
Hamilton Books Acquisitions Department (301) 459-3366

Estover Road
Plymouth PL6 7PY
United Kingdom

Printed in the United States of America
British Library Cataloging in Publication Information Available

Library of Congress Control Number: 2007922310
ISBN-13: 978-0-7618-3742-8 (paperback : alk. paper)
ISBN-10: 0-7618-3742-6 (paperback : alk. paper)

™ The paper used in this publication meets the minimum requirements of American National Standard for Information Sciences—Permanence of Paper for Printed Library Materials, ANSI Z39.48—1984

This Book Is Dedicated To

The Memory Of

Nancy Kephart Collins

1941–2000

Table of Contents

List of Appendices

List of Figures

List of Tables

Preface

I have had an interest in my family history for a long time but had little time to devote to research while working and helping to raise my own family. However, after retiring and then losing my wife to cancer, I decided it was time to put down in writing my family history for my own children, grandchildren, and future generations.

This book is about one side of my family history—the Van Sickles or with special emphasis on my great-great grandfather Isaac Van Sickle. During the 18th century and early 19th century, there were multiple spellings of the surname. Although I have used the more accepted Van Sickle, one can also find Van Sickel, Sickle, Sickles, Cycle, and VanCycle, among others. A letter from Pastor Joseph Van Sickle, a grandson of David Van Sickle II, dated March 12, 1936, supports the variety of spellings.

This book chronicles Isaac Van Sickle's story from his early days as a farmer with a young family, through his service in the Civil War and, finally, to his later years that were impacted by his service in the Civil War. This story also includes that of the Van Sickles that came to live in western Maryland near the end of the 18th century. This story comprises the participation of Isaac, his younger brother David, his brother-in-law Jefferson Davis, and other members of the Van Sickles clan in the Civil War and other events that transpired during his lifetime. Isaac's story is probably not much different from that of many others who lived during this tumultuous time in 19th century America.

What I thought would only take a short time, turned into a wonderful adventure that has taken me over 2-years to navigate. During my research, I learned much about the events that helped shape their lives and the exciting, yet tumultuous, period of American history in which they lived. I visited many communities, toured battlefields, spent countless hours on research, read a multitude of written material, and met with many wonderful, dedicated, and helpful individuals, making for an exciting and rewarding adventure. I even spent a week taking a seminar on researching old military records at the Sanford University Institute of Genealogy and Historical Research. With my brother, I spent many, many hours going through Civil War company muster rosters and pension records pertaining to the Van Sickles. Hopefully, I have been able to translate this information into a format that is meaningful, historically accurate, and something that will make my family proud of their heritage.

I need to thank a number of people for their help in making this book possible. First my brother, *retired Lieutenant Colonel Harry D. Collins, USA*, for his extensive genealogical research; second, Timothy Reese for his guidance and early editorial assistance; and third, Cathy Beeler of the Monocacy National Battlefield for her editorial comments and review. I also greatly appreciate the assistance of the staff at the Maryland Room of the C. Burr Artz Central Library of Frederick County, the Historical Society of Frederick County, the Friend Family Association of America, the Garrett County Historical Society, and the National Museum of Civil War Medicine.

Special thanks are due to Yvonne Miller for putting together the camera-ready art and other associated information required by the publisher and to Meredith Burall who used her artistic talent to help design the beautiful cover for this book.

Finally, I want to thank the members of my family for their patience and understanding.

I hope you will enjoy the story that unfolds in the following pages. While written for the family, I hope that any non-family member that reads this book might be prompted to learn more about their own family history. I believe that there are many unique and exciting family stories just waiting to be told.

All profits from the sale of this story will be donated to the Nancy Collins Memorial Fund which supports area hospitals in the fight against cancer. The profits will also be used to support the *You Are Not Alone Program* which provides wood rocking chairs to cancer treatment centers for use by cancer patients, their families, and caregivers.

Chapter 1

The western portion of the fledgling American colonies (the western parts of Maryland and Pennsylvania, eastern Virginia) and all of Ohio in the early 18th century, represented a land of unlimited potential. It required strong hardy and adventurous settlers who were prepared to deal with the many physical and mental hardships and the Indians that awaited them. The Van Sickle clan would be part of this wave of settlers leaving America's eastern colonies to accept the challenges of the frontier.

Part of this frontier was today's Garrett County, Maryland, and western Somerset County, Pennsylvania. This region lies within the Appalachian Plateau of the Appalachian Mountains and is defined in the east by the Negro, Meadow, and Big Savage (height of 2,800 feet) mountains and in the west by Laurel Hill Ridge (also known as Laurel Mountain) and is bisected by the scenic and often rugged Youghiogheny River. The landscape of the region is the result of the methodical sculpting of the mountains by the forces created by weather, with help from its waterways. The result was beautiful mountains, table top fertile farmlands, an area rich in timber, coal and iron, thick forests of pine, oak and maple trees, an abundance of wildlife and lush, picturesque valleys and glades all traversed by the Youghiogheny River and its tributaries.

Indians had used the region for centuries as their summer hunting grounds returning to their permanent homes before the cold, snowy winters of the region arrived. The Indians that used the region were the Six Nations Confederacy that consisted of the Iroquios (Mohawk, Oneida, Onondaga, Cayuga, and Seneca) from the north and Tuscarora from the Carolinas. From the east, the Delaware, (also called the Leni Lenape), from the Midwest, the Shawnee (both part of the Algonquian Nation), and the Mengwees (also called the Mingos). The Shawnee, who called Ohio home, were fur traders and considered cruel and treacherous. They actively resisted the intrusion of the settlers by engaging in constant warfare.

The Youghiogheny River begins its 135-mile journey through the mountainous terrain of Maryland and Pennsylvania at Backbone Mountain in extreme eastern West Virginia and the western edge of Garrett County, Maryland. Backbone mountain, with a height of 3,360-feet, is the highest mountain in the region. The Youghiogheny flows north through Garrett County into neighboring Pennsylvania traversing the western portion of Somerset County until finally ending its journey by joining the Monongahela River near modern day McKeesport, Pennsylvania.

The Indians that used the bounty of the region recognized that the Youghiogheny was a unique river because it didn't flow south like other rivers but flowed north. Literally translated from Indian, the word Youghiogheny means "waters that flow in a contrary direction," although another translation says that it means "three main prongs and the main river."

The region's average height of 2,300-feet above sea level has a profound influence on the region's weather. Winters were harsh with cold temperatures and heavy snowfalls, while the summers were short but beautiful. Spring was brought alive by the melting mountain snow that fed the multitude of the region's waterways. The fall season was full of brilliant colors that painted a dazzling picture of Mother Nature at her seasonal best. The region was covered with mature, dense forests that provided an ample supply of building materials and provided homes for a variety of wildlife including fish, deer, and black bear.

The first white men entering the region relied on the existing Indian trails such as the Kittanning path, Nemacolin trail (named for a Delaware Indian chief) and the Turkey Foot trail. The Turkey Foot trail that extended from Fort Cumberland over the Big Savage and Negro Mountains to Confluence, Pennsylvania, was one of the most popular early routes for travelers heading west to the Ohio-Valley. These trails and paths were later to become the wagon roads and turnpikes that would connect the new settlements of the region.

The bountiful wildlife of the region led to a very lucrative, although dangerous, fur trading business between the adventurous early frontiersmen and the Indians. Items such as knives, pots, pans, cloth, trinkets, and, even, firearms were traded to the Indians for a variety of furs including beaver, fox, otter, and mink. The demand for the furs by the wealthy upper class in both the Colonies and Europe was staggering. This, coupled with the need for land by the ever increasing population of the Colonies, resulted in an ever increasing flood of settlers to the western portions of Maryland, Pennsylvania, Ohio, and Virginia, thereby, placing further pressure on the Indians in the regions.

The Ohio Company, a Virginia trading company, was formed in 1749 to trade with the Indians. Their agent, Christopher Gist, was probably one of the first white men to extensively travel the region. Gist made two trips to the regions. The first trip was in 1750 and the second trip was sometime during 1751 and 1755. Gist made these trips to research a portion of the 400,000 acres of land that the Ohio Company controlled. Gist's areas of interest were information on all trails, rivers, mountains, and Indian tribes. Gist's informational base was the Ohio Company's trading post on Will's Creek which was soon to become Fort Cumberland, Maryland. In 1753, the Ohio Company would employ Thomas Cresap to open a route for pack horses from Will's Creek to the Monongahela River.

This was the land (the frontier) that awaited the hardy, adventurous settlers in the beginning of the 1750's. Settlers traveled with their families, goods, equipment, and livestock over the rudimentary road system that connected the east coast with Fort Cumberland (today's Cumberland, Maryland). Fort Cumberland was situated at the confluence of Will's Creek and the Potomac River, and served both as a defense against the area's Indians and their French allies and the jumping off point for settlers heading west.

Fort Cumberland, at that time, had a garrison of 400 men and 10 carriage cannons for heavy firepower. These settlers followed the trails of the early fur traders and Indians into the frontier as the first "wagon road" was probably "built" to support General Edward Braddock in his ill-fated campaign against the French and Indians in 1755. It wouldn't be until 1806 that the Nation's first federally-funded highway, the National Road (today's U.S. Route 40) was authorized by Congress. The engineering required in the building of the National Road was a marvel for its time and included the building of the Casselman River Bridge. Crossing the Youghiogheny River in western Maryland, the 80-foot span was the largest stone arch bridge in America when built and served travelers continuously for over a century. With the construction of this engineered and hard surfaced road, the flow of settlers headed west would become a steady stream that would force the Shawnee and Delaware to go further west to seek other hunting grounds. Construction of the National Road didn't begin until 1811. With the route cut mostly through virgin forests, the road didn't reach the western side of the Allegany mountains and Wheeling, Virginia (now West Virginia), a distance of 132-miles until 1818.

A young surveyor and major in the Virginia colonial militia named George Washington (later to become First President of the United States) made a number of trips through the region from 1753 through 1758. During a visit in 1754, he laid out the Braddock Road that connected Fort Cumberland with a crossing of the Youghiogheny River just south of the Mason-Dixon Line (regarded as the line of demarcation between the north and the south). During the same visit, his Indian scouts noticed that three local waterways—the Youghiogheny, the Casseleman River, and Laurel Hill Creek joined together at what is today Confluence, Pennsylvania—resembled the shape of a turkey foot. This impression was to stick and today that part of Somerset County is called Turkeyfoot Township. During this same period, young Washington surveyed most of what is present day Garrett County.

Several individuals were instrumental in the early settlement of the region. In 1763, Andrew Reams was issued a land warrant for a parcel of land which included a beautiful, fertile valley traversed by Little Laurel Creek. The valley was rich in game and fish and was part of the Shawnee hunting grounds. Reams and a few friends built a stockade on the bank of Little Laurel Creek and from that stockade they were able to defend this fledgling settlement from repeated Shawnee attacks. With the Treaty of Fort Stanwix of November 5, 1768, Reams land warrant became legal. The Treaty of Fort Stanwix was signed by the Iroquois Indians who sold a vast area of land in Pennsylvania to Thomas and Richard Penn for 10,000 English pounds, ending all Indian claims to the land. This authority that had been granted to the Iroquois caused a major rift with the Shawnee and Delaware that resulted in the latter two tribes allying with the French in the French and Indian War (7-years' war) that covered the years from 1756 through 1763. The site of Reams stockade became the Town of Ursina, Pennsylvania, and was open to new settlers. Ursina was named for Judge William J. Baer who owned the land that Reams settled upon. Ursina is the Latin word Ursus which means bear. In the spring of 1770, a large number of mostly German and Irish families departed New Jersey and began the long arduous and challenging trip to Ursina.

In 1765, the three Friend brothers John, Andrew, and Charles migrated from their flood-ravaged farmland in the Potomac Valley region of Virginia to the northern part of today's Garrett County to establish a settlement. The Shawnee Indians had a temporary village on the banks of the Youghiogheny that the Friend's thought would be a great spot for a town. The Friend's were probably surprised (based on the reputation of the Shawnee) when they received permission from the Shawnee to build a log cabin on the site. The next year, John Friend bought the land from the Shawnee. The 159-acre parcel was surveyed June 26, 1776, and identified as "Friends Choice" and was to become the Town of Friend's Fortune which is today's Friendsville, Maryland. Many Amish, German, Irish, and Swiss families soon made Friendsville their home—attracted by the rich farm lands fostered by the Youghiogheny.

Among the early settlers who left New Jersey for the promise of the plentiful land and new opportunities were the five Van Sickle brothers—David (the settler), Ephraim, Isaac, Louis, and Zachariah. Arriving sometime in the 1780's, the Van Sickle brothers were among the earliest settlers in western Maryland in the area of Blooming Rose.

David, the settler, is believed to have initially settled on 100-acres of land that were classified as military lots Numbers 3142 and 3143 and were designated as land for veterans of the Revolutionary War.[1] These 50-acre lots were surveyed for Lord Baltimore's land office and the State of Maryland by Francis Deakins in 1788. Of the 4,165 50-acre lots that Deakins surveyed, 636 lots had been settled, improved, and cultivated by 323 pioneer families. The two lots that David, the settler, settled on were situated on Buffalo Run which joined the Youghiogheny River at nearby Selbysport. By the Act of 1788, those 323 families that had already settled on the lots surveyed by Deakins were able to purchase the lots with a maximum purchase price of 20-shillings per acre. They were further permitted to pay for the land in three equal installments over a 3-year period. On February 26, 1790, David Van Sickle (spelled Vansicle on the contract) entered into a contract to purchase the 100-acres for five schillings per acre.

The boundary between Maryland and Virginia in the late 18th century was very fluid with census records often times showing families residing in Maryland in one census and then in Virginia in the next census, and vice versa.

The history of the original Van Sickle brothers shows that they moved between the two states. A 331$^{1/2}$-acre tract of land surveyed April 25, 1774, called "The Grannary" straddled the Maryland and Virginia state line just west of Blooming Rose. The Grannary was a combination of mature stands of timber and rich fertile farmland. The Grannary is significant because Zachariah Van Sickle lived on the property and had the title of "yeoman." A yeoman was a freeholder who farmed his own land and was considered a man of respectable standing in the community. On February 3, 1829, Zachariah received patented certificate number 1457 from the State of Maryland[2] as he purchased the two military lots (100-acres) from the children of David the settler, Isaac, David II, Catherine, and Elias for $200. David's brother, Louis, reportedly owned the Sickle Hill Farm, located just west of the settlement of Sand Spring, that was also divided by the Maryland and Virginia state line.

Louis Van Sickle was a friend of the famous hunter Meshack Browning. The story is told that in February of 1800, the two teenagers, Louis and Meshack were looking to catch a young deer and make it a pet. They came upon seven deer in a laurel swamp near Blooming Rose that were floundering in the deep winter snow. The now tired, wet, and frustrated boys (who were wearing snowshoes and not about to go home empty handed) caught and killed two bucks using only their knives. During the encounter, Louis was severely bruised, while Meshack came away with only a few scratches. Lure has it that Meshack was such a furious hunter that he killed game with whatever he had—rifle, knife, or club. Meshack spent the bulk of his life in the McHenry area of today's Deep Creek Lake, Garrett County, and had a good friendship with Louis VanSickle. Upon his death, Louis was buried at an unknown site on the Sickle Hill Farm.

That brings us to the final brother David, the settler. He married Mary Little and sometime around December 24, 1796, they had a son named David VanSickle II. The Preston County assessment records of 1798 show that David, the settler, owned one horse, four cows, and two 50-acre lots for a total assessed value of 37 pounds, 13 schillings, and 4 pence. Although the Federal Constitution of 1789 had established a United States currency, some parts of Maryland and adjacent Virginia still used the Colonial Maryland money system which was based on the English currency.[3] The conversion of the pounds, schillings, and pence was equal to $100.52 in United States currency at that time. David II had one older brother, Isaac, and two younger siblings, Elias and Catherine.

Research indicates that David's older brother, Isaac, eventually moved the short distance northward into the Farmington area of Fayette County, Pennsylvania, as both the 1850 and 1860 census place him and his family there. He and his wife, Sarah, had eight children—William, Samuel, Levi, Sarah, Isaac, Ephraim, Amos, and Elizabeth. According to the 1860 census, his older son, Samuel, and his wife also lived in Fayette County. In fact, the two owned adjacent farms and appeared successful, as Isaac valued his farm at $5,000 and personal property at $2,500. Samuel valued his farm at $1,600 and personal property at $800.

David II lived in Preston County, Virginia (now West Virginia), and, in 1816, married Deborah Enlow, the daughter of John Enlow and Elizabeth Frazee. David and Deborah had two children, John who was born in 1817 and Maryanne who was born in 1819. The family moved west to Belmont County, Ohio, shortly after Maryanne's birth. David II joined his brother Isaac, his wife, and two young sons in Archer Township, Belmont County. The brothers began the backbreaking task of clearing the dense forests that had stood untouched for centuries. Once cleared the underlying rich and black soil would provide wonderful farmland. While clearing the forest for farming, David II and Deborah had two more children—Elizabeth who was born in 1821 and Sarah Anne who was born 1823. However, the joy of Sarah Anne's birth quickly turned to sorrow. Deborah developed complications from birth and shortly, thereafter, died. Faced with the responsibility of four young children and the total dedication needed to clear Isaac's land for farming, David II made the decision to return to his original home to help raise his young and motherless family.

Fortunately, David had some help during this difficult time of his life. Grandmother Elizabeth Frazee Enlow had traveled to Ohio to be with her daughter during the birth of her granddaughter Sarah Anne, then returned on horseback through the untamed wilderness to Friendsville, Maryland, carrying six week old Sarah Anne in a basket.[4] Separately, widower David Van Sickle returned to Maryland, and shortly thereafter, married Catherine Harding (also spelled Hardin and Harden). After the deaths of their first two children, Deborah who was born in 1824 and Nancy who was born in 1829, shortly after their births, David and Catherine had five other healthy children. They were Catherine who was born in 1831, Mariah who was born in 1833, great-great grandfather Isaac who was born on March 15, 1835, David Harrison who was born in 1845, and Lydia who was born in 1837. Isaac was born in Preston County, Virginia (now West Virginia). However, the exact location of his birth remains a mystery. Research indicates his birthplace was somewhere in eastern Preston County, in the vicinity of Glade Farms. Although there is the possibility that his birthplace was in the extreme western portion of Garrett County, Maryland, near the community of Blooming Rose. This possibility exists because the state line between Virginia and Maryland was very fluid at that time.

On a hilltop about 2-miles west of Friendsville, an area of 1,100-acres situated on the Blooming Rose Ridge, developed into another community aptly called Blooming Rose, Maryland. It reportedly got its name from the original owner, an Anglican clergyman named Jonathan Boucher. Jonathan, after viewing the abundance of wild flowers growing on the property is quoted by Isaiah, Chapter 35, as saying, "the wilderness and the solitary place shall be glad for them; and the desert shall rejoice, and blossom as the rose."[5] In 1775, Jonathan Boucher could no longer ignore his heart's desire to return to his beloved England, so he resigned his survey for the Blooming Rose land and returned home to England. The property was patented in 1793 and sold by Chapman to a group of land speculators. The speculators divided the tract into lots with many of the lots being sold to slave owning tobacco farmers from southern Maryland. This small community had the first Methodist church and first school in modern day Garrett County. It became a thriving farming community and in the 1790's and early 1800's, reportedly, had a new farm opening every mile or two. This area was known for its rich "black walnut soil" that produced abundant fields of golden wheat with tall corn and oats. The area was bordered by rolling ridges covered with mature stands of giant oak and maple trees. It was at Blooming Rose that Isaac Van Sickle married Louisa Rebecca Davis on September 15, 1856. The marriage was performed by the Reverend Thomas Doll. Today, Blooming Rose consists of a small Methodist church and a large and well kept cemetery. A visit to the church revealed why it was a popular place to settle. The Blooming Rose area has beautiful rolling hilltop farmland and provides a spectacular view of the surrounding countryside.

Another settlement located between Friendsville and Ursina, just south of the Mason-Dixon Line was established by Evan Shelby. He was a captain in the French and Indian War and decided to stay in the area after the war ended. Just south of where Buffalo Run joins the Youghiogheny, he started a trading post on 149-acres of land patented in 1773. With its excellent location on a navigable portion of the Youghiogheny, the trading post soon became the Village of Selbysport, Maryland. In preparing for what he hoped would be an influx of settlers, Selby had his land laid out in lots but the influx of settlers didn't happen. Selby was disappointed and sold the land to the Frazee family from New Jersey and moved to Tennessee. Selbysport never became a significant community like its neighbors but it did have the distinction of being the first town in Garrett County and the site of the county's first grist mill.

Great-great-great grandfather David Van Sickle II and his second wife, Catherine, lived in a number of areas in today's Garrett County before moving a few miles north across the Mason-Dixon Line to Ursina, Pennsylvania. Before their move to Ursina, the 1860 census for Selbysport listed the Van Sickles as residents. David, Catherine, and David H., were one household; Isaac and his wife, Louisa, and their two children Jefferson, age 4, and Susan, age 2, were listed as another household. Isaac listed his occupation as a farm laborer.

In 1840, Isaac's half sister, Sarah Anne Van Sickle, married Joshua Monroe Friend. Joshua was the son of Andrew Friend who was one of the founders of today's Friendsville. This marriage represented a further intertwining of the Van Sickle family with the other pioneering families of western Maryland namely, the Enlows, Frazees, and Friends as Catherine's older sister, Magog Harding, had earlier married Andrew Friend. Their son who was born in 1818 was none other than Joshua Monroe Friend who married Sarah Anne Van Sickle. At this early time in America's history, the Van Sickles had firmly established themselves as one of the early settlers of western Maryland and today's Garrett County.

1. Williams, Judge T.J.C. and Thomas, James W., History of Allegany County Maryland, Volume 1, L.R. Titsworth and Company 1923. Pages 4-9.
2. Maryland State Archives letter dated August 21, 1996 by Ellen V. Alers, Archivist III.
3. Glades Star, Garrett County Historical Society Number 18, June 30, 1945. Pages 140-142.
4. Olsen, Evelyn Guard, Indian Blood. Pages 76-77.
5. Glades Star, Garrett County Historical Society Number 36, December 31, 1949. Pages 378-379.

Chapter 2

The 133rd Pennsylvania

Isaac was a farmer living in Selbysport, Maryland, a small community located in northwestern Allegany County (today's Garrett County), Maryland, approximately 4-miles south of the Mason-Dixon Line. Isaac and his wife Rebecca Louisa Davis (born in Cumberland, Maryland) would later move north to Pennsylvania for better farming.

Isaac Van Sickle stood 5-feet 7-inches tall with a muscular, tanned complexion that had been sculpted by the long hours of hard farming that kept him outdoors in all types of weather. His face was highlighted by his beautiful blue eyes and long brown hair. With this combination of features, Isaac must have been an impressive figure when he put on the Union blue uniform of the 133rd Pennsylvania.

With the beginning of the Civil War and call to arms, many of the wealthy and influential citizens recruited their own military units, commanded by themselves, of course. So George F. Baer, the son of Somerset County, Judge William J. Baer, announced to the citizens of Somerset County that he was forming Company E of the 133rd Regiment Pennsylvania volunteers and he would be captain of the company. Isaac and his brother-in-law, Jefferson Davis, (born in 1835 in Westernport, Maryland) decided to join Company E. Jefferson Davis, his wife, Missouri, and their two young children, James and Kate lived in the Frostburg area when he enlisted in the 133rd. However with his enlistment, Jefferson moved his family across the Mason-Dixon Line to Addison Township, Pennsylvania. To this day, it is not clear whether Isaac or Jefferson caused them both to enlist in the 133rd. We will never know.

On August 6, Isaac walked the 4-miles to nearby Addison, Pennsylvania (location of the Petersburg Toll House of the National Road), to sign his enlistment papers. While there, he mentioned that his wife, Rebecca, was pregnant with their third child and that it was time to harvest the farms' rye crop, and he needed to be home for both events. The enlistment officer said they could delay his call to active duty until the baby was born and the rye crop was harvested. Isaac walked the 4-miles back to Selbysport expecting that he had time but was surprised when he received his orders to report to duty before either event happened. So off to war he went leaving a pregnant wife with their two small children, Jefferson and future great grandmother, Susan, and a field full of rye that needed to be harvested.

Captain Baer's Company E and the 133rd was assigned to the Army of the Potomac's Fifth Corps commanded by Major General Fritz John Porter. It was part of Brigadier General Andrew Humphrey's 3rd Division. The 133rd was commanded by Colonel Franklin Speakman and with three other Pennsylvania regiments (123rd, 131st, and 134th) comprised Colonel Peter Allabach's 2nd Brigade. The 133rd left Harrisburg on August 18 and marched to Arlington Heights, arriving on August 19. Their initial assignment was guard duty in Alexandria, Virginia. On August 30, they were ordered to Fort Ward where they spent 2-weeks alternating between Pickett duty and building entrenchments. On Sunday September 14, the 133rd and the other units—the 2nd Brigade (which now also included the 155th Pennsylvania Regiment) marched 15-miles to Rockville, Maryland. Before continuing, they added to their equipment packs light shelter tents and 60 rounds of ammunition for their newly issued Springfield muskets. They resumed their march over dusty roads under a relentless late September hot sun and arriving in Frederick, Maryland, the afternoon of September 15. Upon reaching Frederick, the fatigued and dust covered troops stopped to enjoy the cool waters of the Monocacy River. Humphrey's 3rd Division including the 133rd Pennsylvania, received orders to remain in Frederick and defend the area against any possible attack from the Confederate troops that were reported to be moving in the direction of Frederick. After a 1-day stay, Humphrey's 3rd Division resumed its march and headed for Antietam Creek in Sharpsburg. As they resumed their march, Isaac Van Sickle and his comrades in the 133rd passed the Union troops that had recently surrendered at Harpers Ferry and were on their way to Parole camp in Annapolis. Little did Isaac know that his brother Harrison and three cousins of Company D of the 3rd Potomac Home Brigade were among those Union troops. Fortunately, the fatigued Pennsylvanians, many with blistered feet, didn't arrive at Antietam Battlefield until approximately 7 a.m. on September 18, the day after the bloody battle. During their march, the citizens of central Maryland and Frederick welcomed the 133rd with open arms—something that helped to lift the spirits of these new soldiers.

Humphrey's 3rd Division consisting of two brigades (containing four regiments each) of green and untried Pennsylvanians, were ordered to relieve Major General George Morell's 1st Division, Fifth Army Corps. The 3rd Division, including the 133rd and Company E, took up defensive positions at the middle (Porter) bridge that carried the Boonsboro Pike across Antietam Creek, about 1-mile east of the Burnside Bridge. After the battle had ended, they were then assigned the duty of helping clear the wounded and dead from Antietam Battlefield.

As a point of interest, Antietam Creek was crossed by three identical single lane, two-arch bridges—the lower (Burnside) bridge, the middle (Porter) bridge, and the upper (Hooker) bridge. They were built over a time span from 1833 to 1858 with a cost of less than $1,700 per bridge. The bridge's design featured unique pier bulwarks that divided the creek's current and, also, broke up winter ice which accounts for the bridge's longevity. The lower (Burnside) bridge was the only one involved in the intensive fighting of September 17 while the middle (Porter) bridge was a staging area for the Union troops held in reserve. The middle bridge has been replaced by a modern two-lane bridge as it is the main highway connecting Sharpsburg with Boonsboro.

Although unable to identify the specific hospital that Isaac Van Sickle was assigned to as a nurse, the location of the 133rd Pennsylvania at the middle bridge would tend to indicate that he could have been assigned to the Newcomer Farm or Ecker House hospitals (both within a quarter mile of the middle bridge) or possibly 1-mile east at the Pry House hospital on the Boonsboro Pike.

One can only imagine the horrific sight that greeted Isaac, Jefferson, and the other new soldiers of the 133rd. The battlefield was covered with thousands of wounded, dying, and dead soldiers that were intermingled with large numbers of dead animals. One civilian eye witness, John Walter of Frederick, reported that dead and decomposing soldiers and horses were stacked in heaps and that some heaps were stacked 3- to 5-men high and the stench that filled the air was exceedingly offensive. The shattered limbs and mangled bodies were mostly the direct result of the "minie" ball (a conical lead slug more than a half-inch in diameter and weighing over an ounce) that with its spinning motion after being fired from a rifle, destroyed flesh and bones on contact. It is estimated that 93 percent of wounds caused during the Civil War were the result of muskets and pistols.

One must understand that medical care in the early 1860's wasn't prepared for the magnitude and severity of injuries that would result from modern weapons used at close range by the soldiers on both sides. The idea of sterile medical instruments and cleanliness of wounds, bandages, and hospital personnel had not been developed at this time. Each regiment (1,000 men at full strength) had three ambulances pulled by a team of two horses with a stretcher capacity of two to four men. The ambulances could carry six to eight slightly wounded men. Each ambulance had a driver with a sergeant in charge of the group. Stretcher bearers were privates pulled from the ranks of the regiment so that if all ambulances were in use, it wasn't unusual for 12 or more of the regiment's soldiers to become noncombatants. When you add the casualties to this number, a regiment's strength could be severely reduced when needed most—in combat.

Fall came early to Central Maryland in September 1862 with the trees on South Mountain already in their brilliant fall colors and the cold nights providing a glimpse of the brutal winter to follow. The days began with misty mornings with the mist burning off as the warm sun climbed higher in the late summer sky and the daytime temperatures usually only rising into the low- to mid-60's. The first snowfall was on November 7. At the other end of the calendar on St. Patrick's Day 1863, the temperature fell to 14-degrees and there was a mix of snow and rain that turned the earth to a muddy paste. Frederick had snow on March 31, 1863, and, because of the bad winter, no gardens had been planted.[6]

Isaac and his brother-in-law, Jefferson Davis, began their task of clearing the battlefield. Both their physical and mental limits must have been pushed by the demands of the work with the handling of the dead, the horrible stench of the dead, shortage of water, and overall poor sanitary conditions. The protective masks worn by Isaac and Jefferson offered little relief from the horrid smell of the battlefield. It took a week of back breaking work to clear the battlefield with the men getting little rest and exposed to whatever diseases might be in the air.

Isaac was assigned to the brigade hospital. The term "hospital" isn't to be confused with today's modern health facilities as after the Battle of Antietam, churches, private homes, storefronts, barns and sheds were mostly converted to hospitals. Facilities and trained personnel for the treatment, housing, and care of the thousands of casualties from Antietam and the prior day battle at South Mountain didn't exist. Surgeons were in short supply and the vast majority of army nurses had no medical training. They were just ordinary men who were pressed into service because of a need.

The duties of the "army nurses" were, generally, to maintain order among the patients, change linens and bandages, help distribute and assist patients with meals, and any other duties that might be assigned by the surgeons or medical stewards.

Isaac spent the month of October working at the brigade hospital helping to care for the needs of the wounded which included many men who had an arm or leg amputated. It is estimated that as many as 80 percent of the operations performed during the Civil War were amputations. Conditions in the hospitals were difficult for both the caregivers and the patients. Conditions were very crowded, infections and illnesses ran rampant from the unsanitary conditions that prevailed, the quality of food was marginal, water was often impure, medicines were limited, and the, mostly, untrained medical "staff" was grossly overworked. It didn't take long for nurses to find themselves patients.

It was estimated that several weeks after the battle, as many as 100 ambulances a day, carrying both Union and Confederate soldiers would travel through Frederick on their way to the hospitals. It took an ambulance 1-day to make the 22-mile roundtrip from Antietam and the South Mountain battlefields to Frederick.

On November 5, Isaac was transferred to Camp B, in what was known as the "low meadow" portion of the Brunner Farm located on Shookstown Road which was on the outskirts of Frederick, as Patient Number 441. He was hospitalized with debility which is defined as a general weakness, feebleness, and faintness that he probably developed as the result of helping to clear the Antietam battlefield, caring for the wounded, and exposure to the unsanitary conditions at the brigade hospital. However, Isaac would soon find himself, once again, pressed into duty as a nurse at Camp B.

Camp B was one of the two tent hospitals set up on the outskirts of Frederick used principally to house wounded convalescents.[7] It was, in reality, a tent city as the site was covered with dozens of tents. There were tents to house the patients, surgeons, stewards and nurses, kitchens, supplies, etc. The standard canvas tent used for patients was 14-feet by 14-feet 6-inches with a center height of 11-feet and side walls measuring 4-feet 6-inches. The tents had center poles that were held taught by ropes attached to stakes driven in the ground. Each tent could hold 20 patients and was heated by centrally located pot-bellied stoves that were fired by either coal or wood. It wasn't until the first week in January 1863 that Camp B had its own water supply which was an artesian well that had a depth of 136-feet.

Camp B had a capacity of 456 patients although at times it may have had as many as 800 patients. During its 4-months of operation, Camp B handled a total of 1,395 patients. Staff was comprised of five surgeons (a total of 15 saw duty with some spending as little as a week), four hospital stewards (trained medical personnel), 68 male nurses, and 19 cooks. In addition, there were unknown number of area residents who volunteered their services, food, and other items to patients. One of the items supplied by the area residents was underclothing which was in very short supply. This shortage was the result of patients being admitted for medical treatment and recuperation—their uniforms were taken away and they were issued a white shirt to wear and no underclothing.

Surgeon Thomas Reed was in charge of organizing Camp B and its initial month of operation. He was transferred and replaced by Assistant Surgeon Thomas McKenzie on November 21 who remained in command until it closed at the end of February 1863 with all remaining patients transferred to Baltimore, Maryland. During his short tenure at Camp B, Reed complained that he didn't have a hospital steward "capable of dispensing medicines" and had to retain convalescents as nurses because of the utter failure of the military to provide adequate personnel to attend to the convalescents and act as police for the Camp. It is difficult to fault the military for Reed's complaints as no one anticipated the magnitude and severity of casualties resulting from the Battle of Antietam. At the same time, much credit is deserved by Isaac Van Sickle and his comrades that served as nurses. They performed their duty admirably while exposing themselves to disease, infection, abuse, and physical exhaustion.

Things at Camp B were further complicated by the cold, rainy, muddy, and snowy winter. Local residents said it was a very unpleasant winter with more snow than usual. Just before Camp B was closed, the area was hit with a heavy snowstorm on February 22 that left 10- to 12-inches of snow on the ground.[8] The tents offered little protection against the cold and dampness of the winter while, at the same time, the centrally located stoves couldn't adequately heat the interior of the tents and their occupants. The nurses were responsible for maintaining order among the patients and were constantly breaking up fights among patients that were typically caused by boredom. They also had to keep a head-count of patients as it wasn't uncommon for patients to head for home once darkness descended.

As Isaac was attempting to recover from his illness while performing various nursing duties, Sergeant Jefferson Davis was about to have his first taste of combat. As part of General Burnside's Army of the Potomac, the 133rd left Sharpsburg on October 30 and proceeded to Falmouth, Virginia, which was located across the Rappahannock River from Fredericksburg. Specifically, the 133rd was part of Brigadier General Daniel Butterfield's 5th Army Corps assigned to the 3rd Division commanded by Brigadier General Andrew Humphries and in Colonel Peter Allabach's 2nd Brigade.

The men of the 133rd spent the next month encamped at Falmouth, Virginia, consistently engaged in drilling while awaiting orders. Meanwhile, General Lee's Army of Northern Virginia had established strong defensive positions on the hills (called Marye's Heights) overlooking the town of Fredericksburg. The 133rd and the rest of the Army of the Potomac watched the Confederates prepare for a possible conflict. On November 21, Burnside asked for the surrender of the Town of Fredericksburg and got a resounding "NO" from its mayor. There was little doubt that these two exhausted armies, still recovering from the brutal fighting at Antietam only 2-months before, were headed for another battle.

The early and cold winter that had invaded Maryland didn't spare Virginia or the two armies camped there. The first week of December found heavy snow and subfreezing temperatures that, reportedly, resulted in ice forming inside houses. It is hard to imagine the suffering the soldiers endured while living in tents and awaiting the arrival of the pontoon bridge that would permit the crossing of the Rappahannock and the awaited battle between the two armies. The waiting appeared over, when on December 9, Burnside issued orders that each man be issued 60 rounds of ammunition and 3-days of cooked rations. But delay continued on the building of the pontoon bridge as the Confederates, from their excellent defensive position, were able to harass the bridge builders while the heavy early morning fogs made the coordination of troop movement very difficult. Frustrated by the delays, Burnside attempted another attack on December 12 but was forced to withdraw because of heavy fog.

December 13 dawned with heavy fog but Burnside, again, sent his forces across the pontoon bridge. However, this day was to be different as by 10 a.m. the fog had thinned enough to enable the various units to communicate their movements and the Union attack began. One must understand that awaiting the attackers were 76,000 veteran and confident Confederate troops occupying superior defensive positions that were basically undefeatable. Their defenses were established on high ground with clear fields of fire as Lee was the master of defensive battles and Burnside's Army of the Potomac would soon find out.

After advancing through the town of Fredericksburg, the Union force's first attack on Marye's Heights began at 11:30 a.m. The attackers were met with a virtual sheet of flame from the Confederate defenders and their ranks were quickly devastated. The slaughter of the Union attackers continued throughout the day as wave after wave of Union attackers were unable to proceed further than the sunken road at the bottom of Marye's Heights. Each attacking wave was further slowed by a battlefield covered with wounded and dead comrades, discarded equipment, and brutally cold weather. Finally, sometime between 2 and 3 o'clock in the afternoon, it was time for General Humphries' 3rd Division to cross the river. Although under constant shelling from the Confederate artillery, they successfully completed the river crossing on the newly built pontoon bridge and prepared to attack Marye's Heights. The troops were ordered to unsling their knapsacks, fix bayonets, and not to fire their muskets until on top of the defenders. The experience from previous attacks had shown firing and reloading of their muskets during the charge slowed the charge and made them easy targets.

Humphries' two brigades of untried Pennsylvanians including Jefferson Davis and the 133rd moved out smartly but like the assaults that preceded, they were doomed to failure. Although demonstrating great valor and courage, the 133rd and their valiant fellow Pennsylvanians were stopped only 50-yards from the stonewall atop Marye's Heights by the withering, concentrated fire of the defenders. The 133rd suffered heavy casualties with Jefferson Davis shot in the left leg. Davis stayed on the battlefield with his fellow comrades in the 133rd until ordered to withdraw from their position at about 3 a.m. on Sunday morning, December 14. While the 133rd regrouped and was issued fresh ammunition in the Town of Fredericksburg, Davis was sent to the brigade hospital for treatment. With the withdrawal of Humphries' two brigades, the assaults at Marye's Heights were finished for all practical purposes although the 133rd stayed in Fredericksburg until Tuesday morning before withdrawing back across the Rappahannock River. This battle was a military disaster for General Burnside and his gallant Union troops.

It is hard to imagine what Sergeant Jefferson Davis and his fellow Union comrades endured during the continued attacks on Marye's Heights. The men must have been cold and exhausted. They were probably deafened for hours after the battle from the explosions of artillery, constant rattle of rifle muskets and pistols, commands of officers and noncoms, bugles, swearing, and cries of the wounded. In addition, the air was probably fouled and virtually unbreathable from the heavy sulfur smoke given off by the black powder used in the muskets. The muskets themselves could become hot enough that their metal barrels could burn the hand. Finally, the sight of thick smoke, muskets firing, seeing friends shot, and the sight of mutilated bodies could unnerve even veteran soldiers let alone the untried 133rd. The withdrawing soldiers must have resembled beings from another world with faces streaked with a mix of sweat, black powder and blood, along with lips and teeth black from tearing open the paper cartridges that contained the powder used to charge their muskets.

The night of December 13 was clear and bitterly cold with the Northern Lights shinning brightly on the battlefield revealing the grotesque sight of frozen bodies of the gallant Union soldiers who died in the fruitless attacks on Mayre's Heights. While the Union spent December 14 clearing the battlefield of the wounded who had survived the previous night's exposure to the freezing cold, Jefferson began his journey to recovery. He was sent to the hospital at

Point Lookout in southern Maryland and after treatment was sent to Chestnut Hill Hospital in Philadelphia, Pennsylvania, for recuperation. While at Chestnut Hill, Jefferson was promoted on January 15, 1863, to 1st Sergeant.

The 133rd suffered 20 killed (including three officers) and 137 wounded (including eight officers) plus 27 men missing for a total of 184 men.[9]

Meanwhile, Isaac continued his nursing duties at Camp B until its closing at the end of February 1863. Dr. McKenzie of Camp B was sent to Harpers Ferry to take charge of the hospital and all remaining patients were transferred to Hospital Number 1 situated further down Shookstown Road near Brunner's Mill. So with both Isaac and Jefferson recuperated from their particular medical adventures, they were reunited with their fellow comrades in the 133rd in Falmouth, Virginia. Major General Joseph Hooker had replaced Burnside as Commander of the Army of the Potomac after the debacle at Fredericksburg. The 5th Army Corps had Major General George Meade as its new commander with the 133rd still part of Brigadier General Andrew Humphries' 3rd Division.

By late April, Hooker had assembled an army of 134,000 men across the Rappahannock River from Lee's 63,000-man Army of Northern Virginia. Although underfed and ill-supplied, Lee's army still held the same excellent defensive positions they had used to defeat Burnside's Army of the Potomac at Fredericksburg in December.

While Hooker and Lee played chess with their armies in preparation for the upcoming battle that would take place in Chancellorsville, Virginia, General Meade with the 5th Corps including the 133rd reached Chancellorsville the night of April 29 and was placed in reserve. From May 1 through 4, the two armies fought a bitter battle with Lee's military strategy overwhelming Hooker's superior numbers. Luck finally smiled on the 133rd which had been so badly abused at Fredericksburg. The 5th Army Corps was held in reserve and, although they occupied various defensive positions during the 5-day battle, they scarcely fired a shot. When Hooker withdrew from the field under the cover of heavy rain, the Union had lost another battle and the outnumbered Confederates had again embarrassed the numerically superior and better supplied Union Army of the Potomac.

The 9-month enlistments of Isaac and Jefferson ended and they returned to Harrisburg where they were discharged on May 26, 1863. Unfortunately for Isaac at discharge, he owed the Army $32.51 (almost 2-months pay) for lost uniforms which probably happened when he became ill and was hospitalized.

Isaac returned to his wife, three children, and farming while Jefferson resumed teaching with a slight limp from his wound received at the battle of Fredericksburg. They would remain civilians for less than a year as they couldn't pass up the $300 enlistment bounty offered by Maryland for men willing to join the Potomac Home Brigade while both lamented the fact that they had missed the fierce Gettysburg battle.

6. Quynn, Allen G., personal diary of Weather for Frederick County, Maryland, 1857-1864.
7. Reimer, Terry, One Vast Hospital, National Museum of Civil War Medicine, Pages 99-100, 102.
8. Quynn, Allen G., personal diary of Weather for Frederick County, Maryland, 1857-1864.
9. Bates, History of the Pennsylvania Volunteers, Volume VII, Pages 263-265 and 272-274.

Chapter 3

Before Monocacy

During the Civil War, Maryland Union troops didn't have separate battle flags like many of the regiments from other Union states. Maryland didn't adopt an official state flag until 1904, although a number of unofficial flags were in existence before that time. In fact, Maryland military units in 1888 were using the flag that was later adopted in 1904. For this book, I am using the 1904 flag, which is the only U.S. state flag based on heraldic emblems.

The 3rd Regiment of the Potomac Home Brigade was formed during the period from October 31, 1861 through May 20, 1862, with a service term of 3-years. Companies A, B, C, D, and H were recruited in Allegany County (which included today's Garrett County), Maryland. The 3rd Regiment was originally assigned to the Mountain Department and then was reassigned to the Middle Department and the 8th Corps in June 1862. After the Battle of the Monocacy, the 3rd Regiment became part of the 6th Corps and the Army of the Shenandoah. The 3rd Regiment was reassigned for the last time in August 1864 to the Department of West Virginia and served there until being mustered out May 29, 1865.

The two corps badges in which the 3rd Regiment served are shown below. These badges are the result of the corps identification system instituted by General Joseph Hooker upon taking command of the Army of the Potomac in January 1863.

I believe it's important to provide some background information before discussing the Van Sickles participation in the Civil War. I will attempt to provide a brief overview of the life of a new soldier along with a profile of Maryland.

At the beginning of the Civil War, Maryland was an integral component in the nation's transportation network and economy. It offered multiple modes of transportation for both goods and people from the east coast to the growing west. One could travel across Maryland by highway, rail, or water. The economy of Maryland was also similar to its southern neighbors, depending heavily on farming, livestock, and tobacco.

Maryland's National Road (today's Route 40 and Interstate 70/68) connected the east (Baltimore) with Pittsburgh, Pennsylvania, and the west. The principle vehicle for moving goods was the Conestoga—a wagon which was reportedly developed in the later part of the 18th century by the Mennonites living in the Conestoga Valley of Lancaster County in eastern Pennsylvania. It was a huge, bulky wagon with four oversized wheels, a white canvas top measuring

28-feet in length and was pulled by a team of six horses. It had a load capacity of 5,000 pounds and was ideal for transporting bulk cargos of flour, corn, oats, and whiskey and had the ability to travel 12- to 14-miles per day over all kinds of roads. Travelers reported that it was common to see seven fully loaded Conestoga wagons on each mile of the National Road. The Conestoga wagon reached its "hey day" between 1820 and 1840 when it was affectionately referred to as "the ships of inland commerce."

The Baltimore and Ohio Railroad (B&0) was the east-west rail system connecting Baltimore with Frederick, Harpers Ferry, Cumberland, Wheeling, Pittsburgh, and points west. Started July 4, 1828, as the first United States passenger railroad, the B&0 was still a young, expanding railroad at the beginning of the Civil War. The B&0 was capable of hauling large quantities of both goods and passengers with over 200 locomotives and 3,500 railcars in a shorter time than army wagon trains. For example, a full regiment of 1,000 men with all of its equipment, stores, transports, and animals could be moved in a train consisting of two or three locomotives with tenders and approximately 40 rail cars that could travel up to 15-miles per hour over flat track. Under the leadership of John Garrett, elected President of the B&0 on November 17, 1858, it would play a significant role in the upcoming Civil War for the Union that would hasten the growth of this new form of transportation.

On February 22, 1823, Virginia incorporated the "Potomac Canal Company" which was to create a canal connecting the east coast with the Ohio River in the west. During the next several years, Maryland, Pennsylvania, and the U.S. government further developed Virginia's idea including funding, ownership, criteria for building the canal, and timeline for construction. After 5-years of planning and a multitude of proposals, on July 4, 1828, President John Quincy Adams turned over the first shovel(s) full of dirt that began "The Great National Project"—the building of the Chesapeake & Ohio (C&O) Canal. Originally planned to connect with the Ohio River at Pittsburgh, Pennsylvania, increased costs and construction delays resulted in only 184.5 miles of the canal being built with its terminus in Cumberland, Maryland.

Even in its shortened length, the C&0 Canal was an engineering marvel that was masterminded by an engineer named Benjamin Wright. The canal followed the path of the Potomac River and when completed in 1850 had 11 multi-arch stone aqueducts, a 3,118-foot long brick-lined tunnel, 8 dams, 75 locks with a minimum depth of 6-feet; the top width of the canal was approximately 40-feet.[10] The jewel of the C&0 Canal was the mammoth Monocacy Aqueduct which carried the canal over the Monocacy River at its juncture with the Potomac River. The Aqueduct was a seven-arch structure that spanned 560-feet and was constructed of pink quartzite (a type of granite) that was extremely strong and durable.[11] During the Civil War, the Confederates attempted a number of times to destroy the Aqueduct but its construction with the granite made it virtually indestructible. With approximately 500 canal boats each capable of carrying 120-tons of coal from the mines in western Maryland, the canal supplied the hungry iron furnaces of Baltimore. In addition to coal, it also transported grains and lumber to satisfy the needs of the east coast.

However, the importance of the C&0 in transporting goods would be short lived. The B&0 would soon demonstrate its ability to move greater amounts of goods and larger numbers of military personnel faster and to more locations than the C&0. These factors, coupled with the continued disruption of the canal's operations by ongoing Confederate raids and the canal's vulnerability to the flooding of the Potomac River, helped in the canal's decline in importance and eventual closure.

Not to be overlooked was the fact that all goods moving southward had to pass through Baltimore. By road, it was the old Post Road (today Interstate 95) and by rail it was the B&0 which was just south of Washington, D.C., and connected with the Orange and Alexandria Railroad which serviced Virginia and the Confederate capital, Richmond, Virginia.

Baltimore was the center of manufacturing and commerce for Maryland and the region. It produced significant quantities of iron, clothing, boots, and shoes—all products required by the military. It was also a major shipping port (with access through the Chesapeake Bay) and carried on a lively trade with South America. Baltimore docks unloaded coffee, copper, and bananas while shipping grain, corn, and beef.

Baltimore was the largest industrial center south of the Mason-Dixon Line and had a large population sympathetic to the south.

In the 1860 census, Maryland's official population was 645,000. Maryland also had a slave population of 87,000 but that number wasn't included in the census numbers. Baltimore comprised 40 percent of the state's populace while only 15 percent lived in the 3 western counties of Allegany, Frederick, and Washington. Occupation makeup was similar to its southern neighbors with 29 percent in farming, 21 percent as laborers and 12 percent servants. Overall, more than 60 percent of Maryland's workforce was non-skilled.

Maryland's border with Pennsylvania was defined by the imaginary Mason-Dixon Line (surveyed by Charles Mason and Jeremiah Dixon from 1763 through 1767) while its border with Virginia was defined by nature in the form of the Potomac River. From a military prospective, the Potomac River provided a natural defensive barrier against the Confederate armies if Maryland was a member of the Union.

South of Maryland and the Potomac River was the beautiful Shenandoah Valley. The Valley is part of Virginia and is bordered on the east by the radiant Blue Ridge Mountains and to the west by the grand Allegany Mountains. The Valley is 140-miles long and 24-miles wide in the north tapering to only 12-miles wide in the south. Crafted by the Shenandoah River and its two main tributaries, the South Fork and North Fork Rivers, the valley is home to a bountiful and fertile farmland.

The Valley was the breadbasket of the Army of Northern Virginia. It boasted one of the longest macadem surfaced, graded, all-weather roads in

North America called the Valley Pike (basically the route of today's Interstate Route 81). This roadway wasn't only a vital means of transporting the valley's goods and products to market but also was invaluable as an invasion route to the north by the Confederates. It provided a route of unimpeded withdrawal and an area for rest and recuperation for the Confederates after engagements with the Union Army. It also provided a great staging area for Confederate sorties into the north and represented a constant backdoor threat to the Union Capital—Washington, D.C. These are some of the main reasons why the Valley was to play such an important part in the Civil War and the fighting that would plague central Maryland throughout the war.

On March 6, 1861, faced with the growing threat of war, Congress passed legislation creating an army of 100,000 men who had a service term of 1-year. This single army representing all of the northern states was a departure from the previous method of individual states supplying their militias to meet the majority of the military needs of the nation. At this time, the United States Regular Army consisted of approximately 16,000 men including 1,098 officers with most assigned to duty in the west. In 1861, Maryland had virtually no state militia and many of its sons were crossing the Potomac to join the Confederate Army.

Soon after passage of the March 6, 1861, legislation, Maryland Governor Hicks issued a call for volunteers to form four regiments for Maryland. The regiments were named the Potomac Home Brigade and their duty was to protect the B&O Railroad track, all bridges, and important buildings. They were also to help control southern sympathizers in Baltimore if the need should arise.

Those in political control in the north recognized the importance of Maryland to the Union due to its transportation, economics, location, and the Union Capital of Washington, D.C. They recognized the strong southern sympathies of Marylanders and the real possibility that Maryland could join the Confederacy. In the 1860 Presidential election, the southern Democratic candidate John Breckinridge received slightly over 42,000 votes versus the less than 2,300 votes received by Republican President elect Abraham Lincoln. Taking all factors into consideration, the Union took the bull by the horns and ordered troops into Maryland to protect Washington and to guaranty Maryland's allegiance to the Union. Under this pressure, Maryland moved the vote of its state legislature from Annapolis to Frederick to effectively eliminate the influence of southern Maryland delegates sympathetic to the Confederacy and secession, and guaranty its vote to remain in the Union. With this action, Maryland became a member of the Union (a border state) although many of its citizens would continue to support the south throughout the Civil War. Baltimore was to become a center of intrigue and spies, while many even considered it the northern most city of the Confederacy.

After the Union debacle at Bull Run on July 21, 1861, the Secretary of War, Simon Cameron, issued the following letter to the Honorable Francis Thomas, member of Congress from the Allegany District. "Maryland and Virginia Home

Guards, it will be perceived that an opportunity is afforded to the loyal citizens of western Maryland, and of the counties of Loudoun, Jefferson, Berkley, Morgan, Hardy and Hampshire, in the State of Virginia, to organize four regiments as a Home Brigade, under lawful authority, for protection of their persons and property from the merciless depredations to which both have been too long exposed." The letter continued to say, "the Home Brigade would be sustained in the vicinity (where enlisted) whilst in service and it would be discreditable to the good and true men of western Maryland, if they failed to avail themselves of this opportunity to defend their own houses and firesides. The Home Brigade will be on the same footing as troops of the regular army, with like pay, rations, equipment . . . " The letter went on to say, " . . . anyone desiring detailed information on this subject, must seek it by letter addressed to ex-Governor Thomas, at Frederick, Maryland, where an officer of the United States Army would be stationed to furnish the required information, and be able to muster into service and arm each company as offered."

With patriotism running rampant, a promise of glory and believing the war would last only a few months, volunteers flocked to the Union enlistment stations, not realizing the years of hardship and horror they were about to face. With the majority of new soldiers, farmers, and laborers, a private's pay of $13 per month was more than most of them made in civilian life. The Van Sickle boys were no different from their fellow farmers and laborers and they too eagerly joined the army.

The 3rd Potomac Home Brigade was organized between October 1861 and May 1862 in Cumberland, Hagerstown, and Baltimore with the majority of the volunteers coming from Allegany County. The actual place of enlistment for the Van Sickles was Nethy Bridge in Allegany County. Ephraim (age 15) enlisted on April 29, Lewis and Samuel (age 31 and 17 respectively) on May 14 and David (Isaac's brother age 17) on June 23. Joining the four Van Sickles in the 3rd Potomac Home Brigade on May 23, 1862, was Harrison (Harry) Glover. Glover had married Lydia Van Sickle in 1857 and, therefore, was the brother in law of Isaac Van Sickle. Glover left behind his young wife Lydia and three small children all under the age of 5. The Van Sickles and Glover were assigned to Captain Michael Fallon's Company D. With many regiments and companies consisting of residents from a single town or community, it isn't surprising to discover that Company D had on its roster eight members of the Friend family, six from the Lewis family, four each from the Glover and Sines families, in addition to the five Van Sickles. Michael Fallon, a 28-year old Irishman, was a resident of Oakland, Maryland, and served as secretary at the mass meeting of the people living in Districts 1, 10, and 15 of southern Maryland, which was held January 11, 1861, and adopted a resolution supporting the Union. Governor Hicks appointed Charles Gilpin as Commissioner of Enrollment for Allegany County and completed the assignment in early September 1861 after enrolling 4,714 soldiers for Maryland's Potomac Home Brigade. Gilpin, a successful merchant from Flintstone, Maryland, was 49-years old and a husband and father of four. He was to become the beloved commanding officer of the 3rd Potomac Home Brigade.

General Lew Wallace would later describe Charles Gilpin as a man who won my confidence at sight. He was of middle-age, had a quiet manner, veteran-ish complexion, iron-grey hair and the evident pride he took in his command. He reminded me as the "Father of the Regiment."[12]

The volunteers then went through a cursory physical examination of their eyes, teeth, jumping up and down, and thumping various parts of the body, all done while naked. They signed the company roll and provided information as to height, color of eyes, hair color, complexion, and occupation. Finally, they underwent a critical physical exam given by a military doctor to determine the fitness of the enlistee for military service.

Statistics compiled from the enlistees show the average Union soldier was 5-feet 6-inches tall and weighed 140 pounds. He was younger than 26-years old with over 60 percent having a fair complexion, just under 50 percent having blue eyes and no dominant hair color. The average age of their officers was just over 30-years. Occupation mirrored the economics of the time with almost 50 percent of the enlistees being farmers.

Isaac Van Sickle didn't follow the rest of the family into the 3rd Brigade, instead he followed his brother-in-law Jefferson Davis into Pennsylvania to enlist in the 133rd Regiment Pennsylvania Volunteers on August 14, 1862. He was assigned to Company E to join Corporal Davis. Isaac was 5-feet 7-inches tall, had blue eyes, brown hair, and a fair complexion. Isaac listed his occupation as farmer and Davis was a school teacher. On January 1, 1863, Davis was promoted to sergeant. Both enlisted for 9-month terms.

The 133rd was organized in August 1862 as part of the state's response to the Call to Arms by Congress. The 133rd quickly filled its ranks and was shipped to Washington, D.C., on August 19 and then to Rockville, Maryland on September 2. The 133rd was assigned to the 2nd Brigade, 3rd Division, 5th Army Corps, Army of the Potomac. We discussed the story of Isaac and Jefferson and their adventures in the 133rd in Chapter 2.

Having successfully passed the military physical, the new soldiers were issued uniforms. While some state regiments had their own uniforms (some Confederate units had blue uniforms which in the heat of battle would result in confusion and mistaken identities), Maryland wasn't one of them. Therefore, the 3rd Potomac Home Brigade was issued standard Union wool blue although they were allowed to add a belt buckle displaying the Maryland State Coat of Arms.

The initial uniform allotment consisted of three each of trousers, shirts, and drawers; two each of coats, pairs of socks, caps, and shoes; one great coat; and other necessary items. To compliment clothing, they also received woolen and rubberized blankets, a knapsack, haversack, canteen, and mess gear. For combat, they received a musket (probably a muzzle loading, smooth bore model 1842) bayonet and accessories. It should be noted that although the smoothbore musket had a reported range of 250-yards, it was only effective at close ranges of

80-yards or less. It is estimated that only 5 percent of the minie balls fired from this type of musket actually hit their target. The introduction of rifled muskets in late 1863 resulted in a deadlier infantry weapon which had an effective range of 300-to 400-yards. This increased range would result in a significant change in infantry tactics in future warfare. The initial total cost to cloth and equip a Union soldier was approximately $80.

The newly outfitted 3rd Potomac Home Brigade then began its rude adjustment to military life. Their days were to be controlled by the sharp blare of the bugle; "reveille" at 5:00 a.m. to begin the day, "taps" at 8:30 p.m. signaling lights out and for all significant activities in between. Roll call was held twice a day to keep track of the soldiers. After morning roll call, the soldiers began their daily activities with some unloading supply wagons, chopping wood, and cleaning up after horses with the remaining soldiers learning how to march. The 3rd Potomac Home Brigade spent the majority of their time drilling in formation with little attention paid to the proper mechanics of loading, aiming, and firing their single shot muskets.

Soldiers were given time in the afternoon for personal matters such as laundry, letter writing, and keeping their personal area and belongings in order for the weekly inspection. During the summer, the men shared small two-man tents that were nicknamed "dog tents." In the winter, the more inventive men would build temporary cabins or set tents on a wooden foundation to help combat the cold, wet winter weather. At best, most soldiers only had their rubberized blanket as insulation between themselves and the hard, cold ground.

Sanitary conditions were terrible with the quality of drinking water questionable at best. Everyone had some form of intestinal illness and lice. When soldiers were issued new uniforms, they usually burned their old uniforms because of the lice infestation. One of the favorite camp past times for soldiers was to have lice races.

Initially, the 3rd Potomac Home Brigade was assigned to the Railroad District of the Mountain Department under the command of the aged Major General J.C. Fremont. Major General Fremont lost the 1856 Presidential election to bachelor James Buchanan. The 3rd Potomac Home Brigade's assignment was to guard the track, bridges, and rolling stock of the B&O Railroad in the upper Potomac region of Maryland and Virginia. During March 1862, the 3rd Potomac Home Brigade was busy. They repaired a wire bridge over the South Branch of the Potomac River, built a telegraph line, and constructed a ferry across the South Branch near Romney, (West) Virginia.

While still recruiting Volunteers to fill its ranks, the 3rd Potomac Home Brigade saw its first action on April 23, 1862, at a skirmish at Grass Lick, Virginia, which is located between the Lost and Coccapon Rivers. A small number of the 3rd Potomac Home Brigade were making a reconnaissance of the area to ascertain the whereabouts of Confederate forces when they were attacked by rebel guerrillas. After a brief engagement centered around the home of Peter Palling, the guerrillas were driven off with light casualties. Under orders from the 3rd's Colonel Downey, the home was burned to the ground.

The 3rd Potomac Home Brigade was then ordered to Wardensville, Virginia, to investigate reports that rebel guerrillas, under the command of Captain John Umbaugh, had murdered a number of Union convalescents near Moorefield, Virginia. After a fatiguing march of 30-miles, the Van Sickles, Glover, and other soldiers of the 3rd Potomac Home Brigade arrived in Wardensville the morning of May 7. Although exhausted, they found the energy to march another 12-miles up the North River and surprised Umbaugh's guerrillas while they were encamped at the home of John T. Wilson. After a brief skirmish, Umbaugh and several of his men were killed and the rest were taken prisoner. With the elimination of Umbaugh's guerrillas, the harassment of Union troops and terrorizing of the local populace stopped. In his report of the action, Colonel Downey said his men deserved the highest commendation for their spirit and vallant action during the skirmish and the hard, ongoing marches to accomplish their assignment.

The 3rd Potomac Home Brigade also saw limited action at the skirmishes in the vicinity of Franklin, Virginia, as Union forces continued their clean-up of the rebel guerrillas (also referred to as bushwackers) operating in the area. The 3rd Potomac Home Brigade returned to Kearneysville, (West) Virginia, and continued to guard the railroad, happy to be out of the rigors and hardship of campaigning in the valley. Little did they know that they would soon find themselves in another battle with Stonewall Jackson that wouldn't end in their favor.

During the same time, Confederate General Stonewall Jackson and his "foot cavalry" had started their march north through the Shenandoah Valley. His troops were called "foot cavalry" because of their ability to repeatedly cover great distances on foot in a very short period of time. During Jackson's Valley Campaign, in 48-marching-days, his troops covered 679-miles. On May 23 at Front Royal, Union forces, under the command of Major General Jonathan Banks (part of Fremont's command), held briefly against Jackson's superior force (16,500 versus 9,000) before hastily withdrawing north to Winchester. At Winchester on May 25, part of Bank's forces, under the command of Colonel J.R. Kenly, made a gallant stand before being overwhelmed by a combination of Jackson's infantry, cavalry, and the wounding of Kenly. In the hasty withdrawal of Bank's forces northward, he left behind a huge cache of supplies, equipment, and munitions for Jackson's army. At the same time, the Confederates cut the Union's east-west communications resulting in much confusion in Washington. With Jackson's success in the Valley, President Lincoln ordered General McDowell to take 20,000 troops and join General Fremont's troops in a pincher movement designed to trap and defeat Jackson's troops and secure the rich farmlands of the Shenandoah Valley for the Union. Stonewall Jackson, receiving word of the Union reinforcements, recognized the Union plan and, in early June, withdrew his 15,000 troops southward. He utilized the captured Union trains for the withdrawal of all of the captured goods they had accumulated and rejoined General Robert E. Lee in Richmond. As a result of Jackson's strategic withdrawal, McDowell and Fremont ended up holding an empty bag after having been thoroughly embarrassed by Jackson in five separate engagements. Jackson's successful Valley Campaign earned him a reputation as an outstanding military tactician.

After being embarrassed by Stonewall Jackson in the Valley, on June 27 the aged and veteran Major General J.C. Fremont resigned his command and was replaced by General John Pope. Pope quickly declared that any male in Northern Virginia who refused to take the oath of allegiance to the Union would be sent south and, if they returned, would be considered a spy. This statement would not sit well in the south and would later come back to haunt Pope.

At this early stage in the war, neither side had made provisions to handle the massive job of housing, feeding, and guarding prisoners. On July 22, a cartel for the exchange of prisoners was signed by both parties. Still believing that the war would be short-term, both sides thought this was a practical solution with released prisoners taking an oath not to take up arms again. The oath was something that both sides didn't honor as soldiers were a commodity in constant need.

Again, I must pause in the story of the Van Sickles in the Civil War to tell you a little about the town in which they were garrisoned—Harpers Ferry.

Harpers Ferry is situated in a beautiful valley created by the Shenandoah River where it joins the Potomac River under the watchful eye of the beautiful Blue Ridge and Elk Ridge Mountains. For centuries, the rich river bottom lands of these two rivers provided bountiful hunting grounds for the regions Native Americans. It wasn't until a ferry crossing was established in 1733 that Harpers Ferry became a permanent settlement. With its excellent location, Harpers Ferry quickly grew into a flourishing community. The natural beauty of the valley and its rivers was so stunning that after visiting Harpers Ferry in 1783 Thomas Jefferson stated, "...this scene is worth a voyage across the Atlantic."[13]

Before the start of the Civil War, Harpers Ferry, located in what was Northern Virginia, was a bustling economic and transportation center for the region with a population of approximately 3,000. The two rivers provided an unlimited supply of water power for manufacturing while the B&0 used Harpers Ferry as its western transportation hub. The Winchester & Potomac Railroad, serving the upper Shenandoah Valley, connected with the B&0 in Harpers Ferry provided the Valley access to eastern markets. Harpers Ferry was also an important stop for the C&O Canal. But the crowning jewel of Harpers Ferry was its armory (one of the two designated by George Washington as munitions manufacturing facilities for the United States). At its peak, the armory provided 100,000 firearms annually.

Harpers Ferry's location in Northern Virginia, separated from Maryland by the Potomac River and its importance as a railroad transportation hub, would prove to be the cause of its downfall when the Civil War came to town.

First, it was the ill-fated raid of John Brown and his rabid followers on October 16 and 17, 1859, that broke the tranquility bringing gunfire and death.

The uprising was put down by two young U.S. Army cavalry officers, Lieutenants Robert E. Lee and J.E.B. Stuart leading a detachment of U.S. marines from Washington. They would later play major parts for the Confederacy in the upcoming Civil War. John Brown's raid was only a wake up call. It was ex-Virginia Governor Henry Wise who brought the stark reality of violence and destruction of the Civil War to Harpers Ferry on April 18, 1861. Virginia militia, under the command of Colonel Thomas Jackson (later known as Stonewall Jackson) attacked Harpers Ferry laying waste to the town. The small detachment of U.S. Regulars set the armory on fire before withdrawing but quick work by Jackson's militia saved most of the valuable manufacturing machinery. The machinery was transported to Richmond and, before long, was producing guns for the Confederate Army.

Jackson remained in Harpers Ferry in control of the important B&O rail junction still hoping that Maryland would leave the Union and join the Confederacy. During this period, Jackson had made a number of demands on John Garrett, President of the B&0, but Garrett refused to honor the demands. The patient Jackson was finally presented with an opportunity to secure much needed locomotives and railcars for the Confederacy. On May 23, 1861, Jackson stopped all rail traffic in both directions on the B&O resulting in the appropriation of 42 locomotives and 386 railcars for the rolling stock-starved Confederacy. These prizes were moved on the Winchester & Potomac Railroad to its end and then Jackson devised a plan to move all 428 pieces over 20-miles of land to the Strasburg and Manassas Gap Railroad. After this successful venture, Jackson demolished everything of value including bridges (both rail and road), rail yards, rolling stock, and arsenal in the Harpers Ferry area. Jackson was also able to transport another 14 locomotives for the Confederacy before leaving Harpers Ferry in mid-June 1862.

In the early summer of 1862, Harpers Ferry could best be described as a war-ravaged wasteland. The once famous armory and town were in ruins and in the opinion of those soldiers garrisoned in Harpers Ferry, the entire place is not actually worth $10.[14] However, the town still had significant value because it was still the transportation hub for east-west commerce and a key supply base for all Union military operations in the Shenandoah Valley. The Union recognized this importance and garrisoned the town throughout the Civil War. It was to become a focal point for Confederate invasions into Maryland and other points north.

August 1862 saw the beginning of a Confederate advance, led by General Robert E. Lee, along with Generals Jackson, Longstreet, and J.E.B. Stuart—the best military commanders the south had to offer. This advance over the next 2-months would create confusion, unrest, and anxiety in the north—ultimately leading to the bloodiest day of fighting of the Civil War.

On August 9 at Cedar Mountain, Virginia, Generals Banks and Jackson once again faced each other. It is difficult to determine which side was victorious, as Bank's troops held their own and Jackson withdrew his forces under the cover of darkness on August 12 to rejoin Lee. A period of posturing followed in which two armies (Pope's 75,000-man Army of Virginia and Lee's 55,000-man Army Of Northern Virginia) took up positions on opposite sides of the Rappahannock River.

Recognizing that his enemy was growing stronger every day, Lee decided that a swift and devastating attack on Pope's supply line was critical. Attacks at Rappahannock Station and Catlett's Bridge were unsuccessful. So Lee turned to his unconventional commander—Stonewall Jackson. Jackson took his 23,000 hungry, threadbare, and shoeless "foot cavalry" on a 2-day, 54-mile march in the summer heat and dust, around the left flank of Pope's army. On August 27, Jackson was poised to attack his objective—Manassas Station.

What Jackson's famished and tired troops found at Manassas Station was overwhelming. Warehouses filled with all types of rations, clothing and ordinance, over 100 new boxcars jammed with similar goods and countless sutler wagons brimming with a wide variety of luxury items, such as cigars, liquor, jackknives, writing paper, and other goods. While Jackson's troops enjoyed the captured Union supplies, all communications between Pope and Washington were severed, leaving both in the dark. The single rail Orange & Alexandria Railroad was damaged stopping rail movement of Union supplies. As Jackson slipped into the surrounding countryside, Pope had no idea were his Confederate adversaries were. Although he did realize that Lee had divided his forces. If Pope could find and attack the divided forces with his superior numbers, he should be able to defeat Lee, thereby, hastening the end of the war. As Pope advanced to Manassas (Bull Run), the night sky was red from the reflection of the burning warehouses, railcars, and countless tons of burning supplies and ordinance at Manassas Station. Whatever Jackson's troops couldn't eat, wear, stuff into their haversacks, or carry in their arms had been set afire. When Pope's forces arrived at Manassas Station, Jackson was gone but had left behind a scene of waste, desolation, and a countryside filled with wanton destruction. In his haste to catch Jackson, Pope had inadvertently placed himself in a poor defensive position if Lee decided to attack.

While the two armies were playing cat and mouse, a Union engineering wizard, Herman Haupt, developed a special construction corps (the forerunner to the World War II Seabees) who soon developed a reputation of rebuilding burned bridges faster than the Confederates could burn them. Four days after the disaster at Manassas Station, Haupt had Pope's critical supply line reopened.

While Pope searched for the elusive Jackson, General McClellan, being an extremely cautious man, refused to send additional troops to support Pope. General McClellan believed the troops were needed to defend Washington from the mounting Confederate threat. As fate would have it, one of Pope's brigades, consisting of regiments from the 2nd, 6th, and 7th Wisconsin, 19th Indiana, and

24th Michigan (later to become known as the Iron Brigade) stumbled upon Jackson's hidden force prompting both sides to prepare for battle.

On August 29, Pope hurled 32,000 Union troops against Jackson's entrenched 22,000 troops with repeated fury but was unable to overrun their lines.

Pope's partner, General Irvin McDowell and his 30,000 troops were totally ineffective in the fighting. While Jackson's partner, General Longstreet kept his relatively fresh 30,000 troops on the sidelines. Pope, thinking Jackson was withdrawing from the field of battle, ordered his forces to attack. Pope's forces were confronted by a determined Confederate force that although exhausted and short on ammunition held their ground. Pope's Army of Virginia fought valiantly. The army continued to give ground until ultimately stopping the counterattacks of the Confederates at twilight.

Pope recognized that he was confronted by a superior force and that his position was precarious and, therefore, began a withdrawal towards Washington. His troops were exhausted and had suffered 23 percent casualties. Although under relentless pressure from the pursuing Confederates, they withdrew in an orderly fashion. Pope was probably saved from further losses and embarrassment by a combination of the weary condition of Jackson's troops and a drenching thunderstorm that made movement all but impossible.

On September 1, 1862, the Second Manassas (Bull Run) campaign ended. Lee's Army of Northern Virginia was only 20-miles from Washington. The Union was in a panic and many wondered whether the Union could win the war since its commanders had displayed both poor leadership and military tactics. Lincoln decided it was time for a change. Pope was sent to Minnesota and his Army of Virginia was folded into the Army of the Potomac and placed under the command of General George B. McClellan.

While Washington was in turmoil, Lee decided to embark on a grandiose plan to invade Maryland. Lee cut the B&0 Railroad (the Union's vital supply line) and, according to some sources even invaded Pennsylvania. He believed McClellan would spend months to organize an offensive. A quick strike into Pennsylvania could further discourage the Union and, possibly, bring recognition of the Confederate states of America by England and France. Lee also believed he could get Maryland to join the Confederacy and when combined with the other attributes of his plan—end the war.

On September 4, 1862, flushed with yet another victory over the Union army, Lee led his hungry, ill-clothed, and exhausted Army of Northern Virginia across the Potomac River at White's Ford located approximately 40-miles northwest of Washington. An estimated 55,000 men started the march. By the time they reached Frederick, Maryland, the morning of September 7, thousands had fallen by the way from sickness, torn feet, and exhaustion. Those who remained were mostly hard-core veterans, subsisting on green corn, green apples, and their loyalty to Robert E. Lee. Upon arriving in Frederick, they anticipated a warm reception complete with plenty of much needed food, clothing, shoes, and new recruits for the undermanned, weary Army of Northern Virginia.

Their reception was just the opposite. Doors were shut, windows shuttered, and businesses closed while there were no volunteers for the army. Lee attempted to remedy the situation by issuing a proclamation to the people of Frederick and Maryland. On September 8, Lee explained they had come to throw off the foreign yoke imposed on the populace by the Federal government and to restore Maryland's "independence and sovereignty." The proclamation did nothing to change the minds of the citizens of Frederick, so the first part of Lee's grandiose plan had failed.

Lee knew that to continue his march to Harrisburg, he would need a dependable supply line for food and ammunition. These supplies would have to come up the Shenandoah Valley via the Winchester & Potomac Railroad to the rail hub at Harpers Ferry. From there, the supplies could then be moved over the B&0 Railroad and C&0 Canal, now controlled by Lee's army. The only thing standing in the way of implementing this was the 12,000 union troops, referred to as the "railroad brigade" that was garrisoned at Harpers Ferry under the command of Colonel Dixon Miles. Miles was considered by many as a "second-rate" commander.

Lee had anticipated that with the approach of his superior military force and the recent defeat of the Union army at Manassas that Colonel Miles would quickly withdraw his outnumbered forces leaving Harpers Ferry undefended. But again, Lee was frustrated. Colonel Miles decided to stay even though he had a very poor defensible position and his railroad brigade that included the 3rd Potomac Home Brigade and the four Van Sickles was clearly no match for the battle-hardened veterans of the Army of Northern Virginia. Another part of Lee's grandiose plan was about to fail.

Lee, therefore, revised his invasion plan and issued Special Order 191 (known as the Lost Order) on September 9; wherein he would divide his Army of Northern Virginia. Dividing one's forces is not considered a sound military tactic but Lee had done that in earlier engagements with success. Jackson took 23,000 troops including Generals Anderson, McLaw, and Walker's divisions to capture Harpers Ferry. General Longstreet would take a similar size force and proceed northwest to Boonsboro, Maryland. Finally, General A.P. Hill would take his 5,000 men and act as rearguard for Jackson.

Lee's plan appeared simple on paper but had to be accomplished in 3-days. Its success depended upon Jackson's exhausted, poorly clothed, mostly shoeless, and short-on-food troops taking circuitous routes to quickly capture Harpers Ferry. On September 10, Jackson divided his troops into three groups. Jackson's campaign-weary "foot cavalry" marched a grueling 51-miles in 2-days (via Boonsboro, Williamsport, and Martinsburg) before taking up a position on School House Ridge just west of Harpers Ferry on the Virginia side. Walker's division occupied Loudoun Heights to the south overlooking Harpers Ferry.

On September 12, Jackson's troops began their assault by shelling the small Union force defending Solomon's Gap. The 3rd Potomac Home Brigade (including the Van Sickle boys and Glover), under the command of Lieutenant

Colonel Downey, along with the Eighth New York Cavalry were defending the Town of Shepardstown, Virginia (West Virginia) when Jackson began his attack. McLaw's division brushed aside the weak Union defenses at Solomon's Gap and continued their advance through beautiful Pleasant Valley. This attack drove the Union defenders back to Maryland Heights which is on the Maryland side of the Potomac River. With the Confederate offensive gaining momentum, the 3rd Potomac Home Brigade was ordered to reinforce the Union troops defending Maryland Heights. At approximately 9 a.m. on September 13, the 3rd Potomac Home Brigade took new defensive positions on the western slope of Maryland Heights. By occupying these strategic sites, the Union hoped to successfully defend Harpers Ferry from the advancing Confederates. Time would soon prove that Maryland Heights would fall into the hands of the enemy and their artillery.

Seizing Maryland Heights by McLaw didn't come without a fight. Kershaw's South Carolina Brigade, of McLaw's Division, descended on the federal position of Maryland Heights after traversing the summit of Elk Ridge. Kershaw's troops were met by four companies of the 3rd Potomac Home Brigade including Company D (the Van Sickles and Glover), commanded by Lieutenant Colonel Downey, and supported by members of the 32nd Ohio. The terrain of Maryland Heights was very rugged and partially covered by dense woods and undergrowth, resulting in very poor vision even on a sunny day. In short, it was terrain that was unsuited for troop movement and combat.

The exchange between the two sides was first brisk and heavy. The minie balls whistling through the trees and hitting the rocky terrain resulted in a cloud of broken tree limbs, rock chips, and dust so thick it was difficult for the fighting troops to see each other. With the Van Sickle boys and their comrades in the 3rd Potomac Home Brigade firmly entrenched behind logs, trees, and rocks, they were able to repulse the initial advances of Kershaw's troops. As skirmishing continued on the Heights, a Confederate force estimated between 8,000 and 10,000 troops was seen advancing on Sandy Hook Road and would soon flank the 4,600 scattered Union troops defending the Heights. At this point, Downey requested re-enforcements if the 3rd Potomac Home Brigade was to hold their position against the advancing Confederate forces. The 3rd Potomac Home Brigade gave ground grudgingly continuing to hold the line until an order to retreat was received. Downey questioned its authenticity and made numerous inquiries as to its origination and authenticity but there was only confusion in the Union chain of command. Indecisiveness appeared to be the order of the day for most of the Union commanders on Maryland Heights. At around 3 p.m., with an exaggerated report that as many as 40,000 Confederate troops were in position to launch an attack on Maryland Heights and the confusion among the Union defenders, led to them retreating across the Potomac River on an existing pontoon bridge. Colonel Downey's 3rd Potomac Home Brigade with the Van Sickles and Glover helped to cover the retreat. Downey said the soldiers of the 3rd Potomac Home Brigade fought well on the Heights and withdrew from their position in an orderly manner when the order to retreat was confirmed. With the loss of Maryland Heights, Harpers Ferry was now at the mercy of Confederate artillery and their superior numbers.

The trap was complete at Harpers Ferry. This was a result of a combination of Jackson's ability to execute Lee's plan, the loyalty of Jackson's exhausted but battle-hardened veterans, the premature order to retreat, and confusion of the Union commanders on Maryland Heights along with Colonel Miles' decision to ignore a report places Jackson and a large force at Boonsboro on September 10.

On September 13, a copy of Lee's "Lost Order" was found by several Union soldiers on the outskirts of Frederick. The Lost Order outlined the disposition of Lee's army, thereby, providing McClellan with critical military intelligence that would result in the two armies meeting on September 17, 1862 at Antietam Creek near Sharpsburg. The battle would become known as the bloodiest single day of fighting in the Civil War.

At Harpers Ferry, Union troops were under a constant artillery bombardment from Maryland and Loudoun Heights with the defenders having no safe place to find shelter. The 3rd Potomac Home Brigade was defending an area on Bolivar Heights extending southeast of the Charlestown Turnpike to the Shenandoah River. Early morning of September 13 found the 3rd Potomac Home Brigade engaged in sharp and sometimes heavy fighting with General Pender's brigade of A.P. Hill's division. The fighting lasted throughout the day, ending at dark with the 3rd Potomac Home Brigade successfully repulsing all attacks. During the battle, Lieutenant Colonel Downey's horse was shot from under him and resulted in the Lieutenant Colonel receiving a head wound. Downey continued to lead his men during the combat to their eventual surrender. The Van Sickles, Glover, and the soldiers of the 3rd Potomac Home Brigade had received their baptism of combat which would prove to be a valuable experience in surviving battles yet to be fought.

Under continued pressure and recognizing their exposed position, a number of Union unit commanders discussed possible escape routes. The first plan was under cover of darkness to cross the Potomac River over the pontoon bridge, follow the path along the C&0 Canal to Harpers Ferry Road, and then follow that road north to safety. This plan was nixed as artillery could not make it up the steep road and the large number of troops couldn't move fast enough to escape detection by McLaws' troops stationed on Maryland Heights. The exposed Union soldiers would be cannon fodder for McLaws' superior firepower. The backup plan was to cross the Shenandoah about one-half-mile south of Harpers Ferry and go south around Walker's division occupying Loudoun Heights and then follow the Potomac to Washington. This plan was also nixed as the crossing was full of deep holes, making it impracticable for artillery and dangerous for the troops. Retreat from Harpers Ferry was impossible but while the main body of the garrison stayed put, a cavalry unit of 1,500 men did manage to escape during the night on September 14 over the pontoon bridge. During their escape, they managed a small parting shoot at the enemy by capturing

Longstreet's reserve ammunition train of 91 wagons near Williamsport, Maryland. The Confederates could ill-afford to lose this ammunition.

Outnumbered by at least two-to-one and weary from the constant shelling which reportedly numbered as many as 50 guns during the final bombardment, Colonel Miles ordered the surrender of Harpers Ferry at approximately 10:00 a.m. on September 15. Brigadier General Julius White actually arranged the formal surrender as Colonel Miles was mortally wounded by a stray mortar round only minutes after raising the white flag.

Jackson had performed another miracle by capturing Harpers Ferry and 12,500 prisoners in only 2-days while also seizing 13,000 arms, 47 pieces of artillery, and a large quantity of ammunition and camp equipment. He had secured Lee's line of supply and communication but the military parole agreement required that Jackson detach Hill's division to arrange for the parole of the Union prisoners who were to proceed to Annapolis for exchange. As a result, Hill's division would arrive late at Antietam when they were needed earlier.

After fighting and under siege for several days, the Union prisoners were permitted 2-days rations and blankets for their march to Parole Camp in Annapolis. They first marched to Frederick where they formed at the Frederick Courthouse before leaving for Annapolis. They arrived in Annapolis on September 22 and totally exhausted. The 3rd Potomac Home Brigade was to find their stay at Parole Camp to be one that would last until May 1863—a period of 8-months. The four Van Sickles, Glover, and their comrades in the 3rd Potomac Home Brigade were to spend many nights during the fall and winter months in the overcrowded Parole Camp sleeping on the ground, drinking contaminated water, and, in general, enduring unsanitary conditions before being returned to duty. They were the lucky ones who didn't fight at Antietam but their antagonists at Harpers Ferry—Jackson, Hill, McLaws, and Walker would all taste the blood of Antietam.

The original Parole Camp was located on the campus of St. John's College in Annapolis and was part of the Department of Annapolis established on April 27, 1861. Parole Camp was the receiving station and a barracks way station for Union troops that had been taken prisoner by the Confederacy and were awaiting exchange. The St. John's College campus contained four buildings (including a former governor's mansion called "Bladen's Foll") and four acres of land adjacent to College Creek and the Severn River. St. John's had approximately 100 students when it was converted to Parole Camp and the mansion was used as College Green Hospital to treat the wounded and seriously ill.

Parole Camp provided the incoming men with medical care, clothing, food, and a place to sleep. In early September 1862, conditions in the crowded camp were reported comfortable with the men having new clothes, fresh bread daily, a varied diet of beef, salt pork, and bean soup plus coffee twice daily. Boredom was a problem, with guards posted to prevent the men from going into Annapolis, Baltimore, and, in many cases, back home.

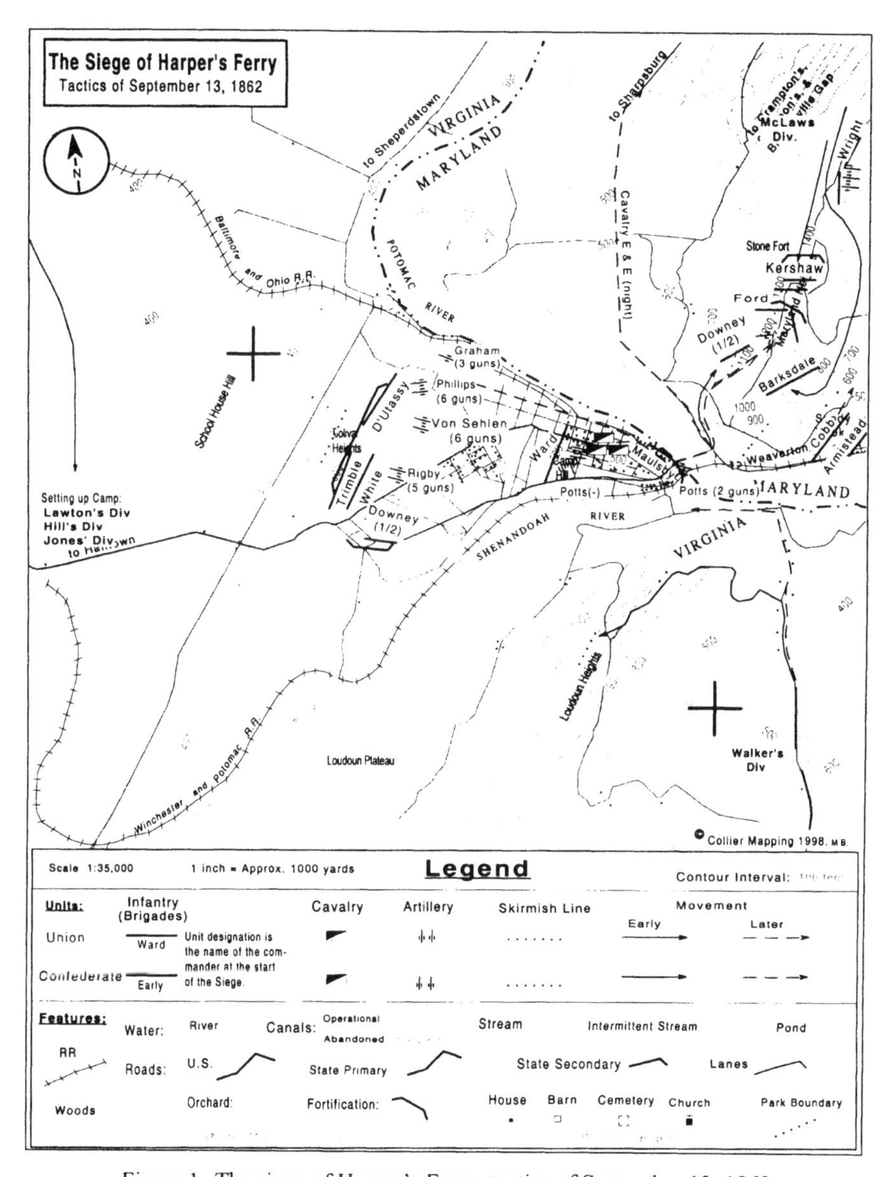

Figure 1. The siege of Harper's Ferry, tactics of September 13, 1862.

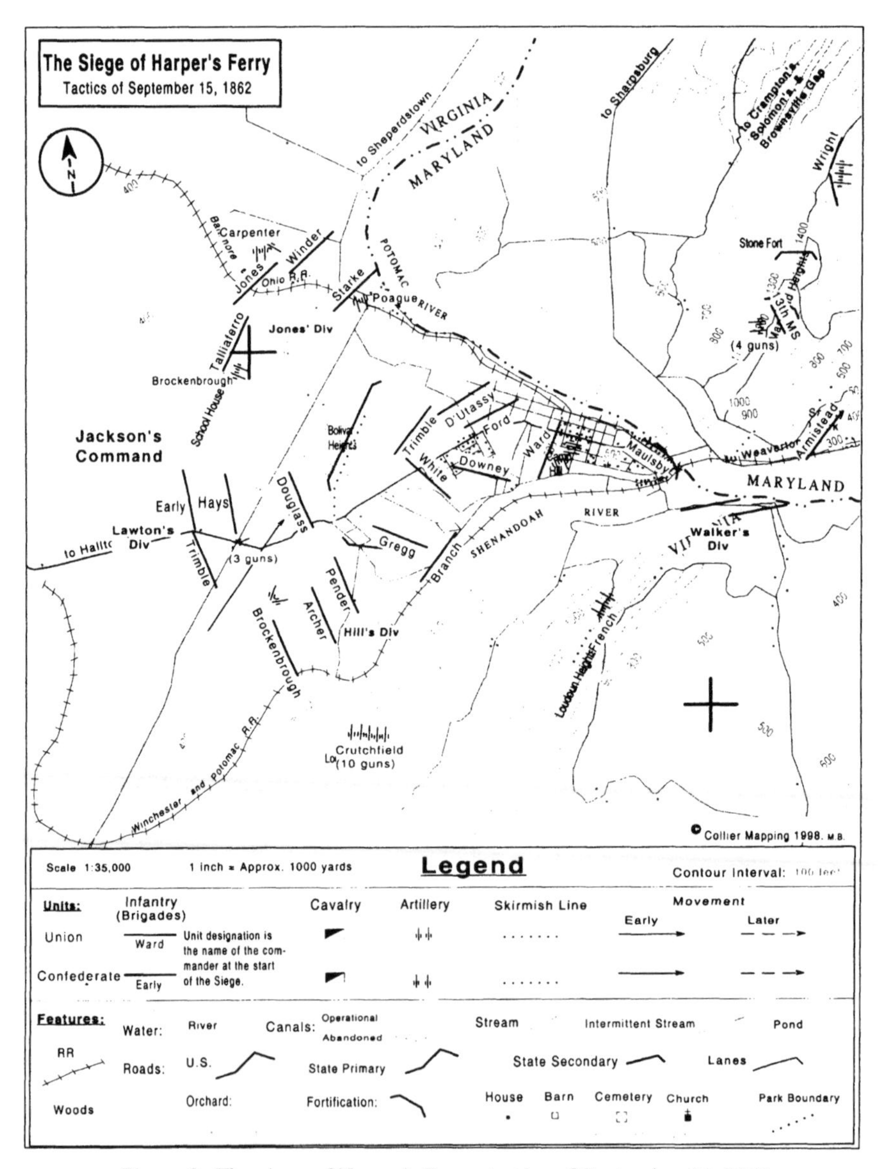

Figure 2. The siege of Harper's Ferry, tactics of September 15, 1862.

With the arrival of over 12,000 Union prisoners from the debacle at Harpers Ferry on September 22, the conditions at Parole Camp were strained to the breaking point with over 20,000 men crowded on the small four acre site. The Van Sickles, Glover, and their fellow comrades were both physically and mentally exhausted from the combat at Harpers Ferry and the march to Annapolis.

Finally recognizing a desperate need for a larger facility, a new Parole Camp was built in 1863 just outside the City of Annapolis. St. John's College didn't receive its campus back until 1866. Today, the area outside Annapolis used for Parole Camp is known as Parole.

Upon its return to duty, the 3rd Potomac Home Brigade, including the Van Sickles and cousin Glover, were assigned the important duty of guarding the vital B&0 Railroad in the area around Elysville, Maryland. The B&0 Railroad bridge, which had been destroyed by the Union during the retreat and engagement at Harpers Ferry, was reopened on September 6, 1863. The reopening of this bridge permitted the vital flow of military goods between Baltimore and Wheeling (West Virginia) to resume. As part of the 8th Army Corps, the 3rd Potomac Home Brigade performed their duty with honor and dignity throughout the remaining months of 1862 and all of 1863.

The Van Sickles didn't escape their early service without side effects from the unsanitary conditions of camp life, poor nutrition, the exhaustion from long marches, rigors of combat, and their extended stay at Parole Camp. Isaac's brother, David Harrison, spent several months in hospitals during 1863 from the effects of diphtheria. Lewis spent December 1862 and September 1863 in the hospital recovering from bronchitis. Both claimed that their illnesses were the result of poor camp conditions and bad water consumed while awaiting parole at Annapolis.

The debacle of the Union garrison at Harpers Ferry became the subject of a military inquiry. In short, testimony determined that the majority of officers serving under Colonel Miles believed that he wasn't competent to command. Reportedly, he was prone to issue conflicting orders and had a habit of countermanding orders he had issued. Major General David Hunter presided over the inquiry and his final report dated November 3, 1862, stated that the evidence indicated that Colonel Miles was unfit to command at Harpers Ferry. The inquiry also indicated that the retreat from Maryland Heights and the surrender of Harpers Ferry were both premature. General McClellan was censored for not moving his troops faster in the pursuit of Lee's invading army. The 3rd Potomac Home Brigade was cited in the final inquiry report as honoring the State of Maryland.[15]

10. Rubin, Mary, Chesapeake and Ohio Canal, Arcadia Publishing, 2003. Page 57.
11. Rubin, Mary, Chesapeake and Ohio Canal, Arcadia Publishing, 2003. Page 51.
12. Wallace, Lew, An Autobiography, Harper and Brothers Publishing, MCMVI, Page 721.
13. Nasby, Dolly, Harpers Ferry, Arcadia Publishing 2004, Page 2.
14. http:///www.nps.gov/hafe/jackson.htm.
15. Official Military Reports of the Civil War, Operations in Northern Virginia, West Virginia, Maryland, and Pennsylvania, Chapter XXXI, Maryland Campaign, Pages 794-800.

Chapter 4

3rd Potomac Home Brigade at Monocacy

Frederick County is situated in central Maryland and is bordered on the west by the South and Catoctin Mountains (part of the Blue Ridge Mountains), on the east by gently rolling lush countryside, to the north by the Mason-Dixon Line, and to the south by the Potomac River. Frederick was known as the "city of clustered spires" as its skyline was dominated by the unique spires reaching skyward from the many churches located in downtown Frederick. It was the center of the region's commerce and the gateway for highway and rail traffic from the west to Washington and Baltimore. The Baltimore Pike (the National Road) the major east-west highway intersected with the primary south highway to Washington, known as both the Washington Pike and the Georgetown Pike in the City of Frederick. Monocacy Junction, also known as Frederick Junction, located just south of Frederick, was a strategic railroad junction of the B&O Railroad and the site of the only iron railroad bridge over the Monocacy River.

The Frederick area was a strong agricultural region with the area populated by neat, well maintained farms that produced an abundance of quality grains, produce, and livestock from the area's fertile soil. With its location so close to Virginia, the loyalties of the Frederick area populace were divided between the Union and the Confederacy.

The Monocacy River is the heart of Frederick County. The Monocacy (the Indian word for meandering) has its headwaters in southern Pennsylvania at the juncture of the Rock and March Creeks joining at the Mason-Dixon Line just north of Taneytown, Maryland. It then meanders through the county, passing just east of the City of Frederick and joining the Potomac River 10-miles south of Frederick for a total length of 87-miles. The Monocacy has a countless number of bends and curves and, being fairly shallow, has many sites that can be forded by both foot and wagon. It is prone to flooding during periods of heavy rain and snow resulting in further nurturing of the fertile farmland. One of the tributaries, Rock Creek, flows through Gettysburg, Pennsylvania, the site of the July 1863 monumental engagement between Generals Lee and Meade.

The Frederick area was no stranger to the horrors of the Civil War. As early as May 1861, an attempt was made to blow up the iron rail bridge over the Monocacy. The Union Military Hospital was established in Frederick in August 1861, thereby, ensuring that Frederick would have a constant reminder of the pain, suffering, and death that followed the soldiers who fought the battles.

In the beginning of September 1862, the Confederate Army under General Stonewall Jackson occupied Frederick. After the bloody Battle of Antietam, over 10,000 wounded and maimed soldiers from both sides were treated in Frederick's six military hospitals.

The 3rd Potomac Home Brigade was formed in early 1862 with a total of eight companies, five of those companies were filled by residents of Allegany County (which included today's Garrett County), Maryland. Most were farmers and laborers that had signed up for 3-years of military service. In July 1864, they had already served over 2-years and were anticipating leaving the rigors of army life and returning to their families and homes. As stated earlier in Chapter 3, the 3rd was commanded by Colonel Charles Gilpin, (a 49-year old merchant and quiet man with iron-grey hair) who before his promotion was commander of Company D. Captain Michael Fallon (a 28-year old single Irishman who had moved west from Baltimore to Oakland) was commander of Company D and the Van Sickle boys at the Battle of Monocacy.

Isaac Van Sickle joined the four other family members in the 3rd Potomac Home Brigade by enlisting on April 11, 1864, for a term of 3-years. This time, Isaac left his wife Rebecca Louisa with their four children—Jefferson, Susan, Ellsworth (born during his service with the 133rd Pennsylvania), and Harry. Isaac was assigned to Company D with his brother David Harrison, and his cousins Lewis, Samuel, and Ephraim, the latter being promoted to Corporal in September 1863. Only 3-weeks earlier on March 23, 1864, Jefferson Davis, his brother-in law also enlisted in the 3rd Potomac Home Brigade for a term of 3-years. Davis was also a veteran of the 133rd Regiment Pennsylvania Volunteers. Davis was assigned to Company I and with his profession as a teacher and prior rank of Sergeant, he was quickly promoted to Company 1st Sergeant as of June 20, 1864.

The 3rd Potomac Home Brigade including Company D was on garrison duty at Annapolis Junction on July 2, 1864, and was part of General Lew Wallace's 8th Army Corps and the Middle Military District. Annapolis Junction was located near Savage, Maryland, and was where the B&O connected with the Annapolis, Washington, and Baltimore railroad that served Maryland's state capital—Annapolis. At that time, the 8th Army Corps consisted of 2,500 mostly inexperienced soldiers with Company D probably having less than 100 effectives. The 8th Army Corps' principal duties were to protect the railroad and control any outburst of the southerners living in Baltimore and the Eastern Shore Region of Maryland. This was not exciting duty but the men got paid monthly and were spared the rigors of long marches and combat. During this time, John W. Garrett, President of the B&O Railroad, advised Wallace that trouble could be on the horizon in Frederick County as some of his railroad employees had reported Confederate raiding parties operating between Harpers Ferry and Cumberland. Opinion was that they probably had encountered Major John Mosby and his 400 men Ranger-Calvary unit that was continually crossing the Potomac to raid Union targets of opportunity.

Suddenly on July 4, the 3rd Potomac Home Brigade was ordered to Monrovia which is located approximately 8-miles east of Monocacy Junction and Frederick. The 3rd was replaced in Monrovia on July 5 by the 11th Maryland and, with new orders, the 3rd Potomac Home Brigade proceeded by rail to Monocacy Junction with each soldier issued 3-days' rations and 100 rounds of ammunition. Little did Isaac, his younger brother David Harrison, his brother-in-law Sergeant Jefferson Davis, and the three other Van Sickles and their comrades in the 3rd Potomac Home Brigade know that they were about to become participants in the battle that saved Washington, D.C. from a Confederate invasion.

The events leading up to the Battle of Monocacy were set in motion on June 13 when Lieutenant General Jubal Early's 2nd Corps of the Army of Northern Virginia departed Richmond, Virginia, to begin the last significant Confederate offensive of the Civil War. He was to clear the Shenandoah Valley of Union forces, invade Maryland, and attack the Union capital of Washington, thereby, forcing General Grant to withdraw troops from the siege of Petersburg, Virginia, to combat the Confederate invasion. At the same time, the Confederacy hoped that the invasion would have an impact on the upcoming Presidential election between the incumbent Lincoln and the challenger General McClellan who was viewed by many as ready to strike a deal to end the war. Early accomplished the first part of his mission by driving Hunter's Army of the Department of West Virginia into the Kanawha Valley and then proceeded to engage Union troops commanded by Brigadier General Franz Sigel. Sigel was from Germany and a school teacher in St. Louis, Missouri, who owed his military command to the substantial political clout he had with the German populace in the Midwest. He was regarded as a poor military commander who seemed unable to win an engagement with the enemy even when having superior numbers. Sigel finally retired from the army in May 1865.

General Early's ragged, mostly shoeless veteran forces had marched for days on dusty roads under the sweltering summer sun from Richmond, Virginia, and through the Shenandoah Valley, a distance of 200-miles. After brushing aside General Hunter's weak resistance, they then defeated General Sigel at Martinsburg, West Virginia, on July 3 gaining considerably much needed supplies as a bonus. Early pursued Sigel as he retreated towards Harpers Ferry, West Virginia. Upon arriving on July 4, Early found that Sigel had withdrawn his troops across the Potomac River into the heavily fortified works at Maryland Heights. Across the Potomac River awaited much needed supplies such as food, shoes, clothing, livestock, forage for their animals, and Yankee greenbacks. But the supplies were unattainable because of Sigel's superior defensive position. Early's forces probably numbered approximately 13,000 men which included 3,000 cavalry[16] and no more than 36 pieces of artillery, supported by a supply wagon train estimated at 400 to 500 wagons. Various Union sources estimated Early had as many as 30,000 men and 125 artillery pieces. With Sigel occupying Maryland Heights, Early was blocked from crossing the Potomac River at Harpers Ferry and the direct route to Washington, forcing Early to take the secondary route that went through Frederick and Monocacy Junction. Before crossing the Potomac at Shepardstown, West Virginia, Early's forces burned the railroad and pontoon bridges that crossed the Potomac River at Harpers Ferry.

As Early's forces continued their march towards Frederick, General Wallace stated that General Sigel reported from his defensive position on Maryland Heights that "he was almost certain that the enemy forces consisted of one corps, three divisions of infantry, and three thousand cavalry. Early, Bradley T. Johnson, John McCausland, Major Generals Robert Ransom, and John D. Imboden are in command."[17]

On July 5, Early's forces crossed the Potomac River near Shepardstown into Maryland on their way to invade Washington. The advance guard for Early's forces was two cavalry brigades under the command of Generals Bradley T. Johnson and John McCausland. They continued to destroy the area's transportation infrastructure by destroying the locks and canal boats of the C&O Canal in Sharpsburg, Maryland. When news of the advancing Confederate forces reached Brigadier General Averill, he withdrew his 3,000 Union troops from Hagerstown leaving the town defenseless. On the night of July 6, a Confederate force of 1,500 entered Hagerstown uncontested. They collected a $20,000 ransom from the town plus a large amount of desperately needed clothing including trousers and an estimated 800 new hats. The total in clothing and merchandise taken from the citizens of Hagerstown was valued at $9,000. The final cost to Hagerstown was covered by the issuance of $39,000 in bonds which took many years for the citizens and the Town of Hagerstown to repay.[18] The Confederates then crossed South Mountain and entered the small, quiet town of Middletown which they pillaged after collecting a meager ransom of $1,500. It was estimated that Early's other plunder from central Maryland included rations for a week for 25,000 men plus 3,000 horses, 5,000 head of cattle, much needed clothing and a mountain of miscellaneous goods. After receiving a shipment of much needed shoes on the morning of July 7, the confident Confederates were only about 5-miles west of Frederick. Confederate General Bradley Johnson, a native of Frederick, had a unit in his brigade called the 1st Maryland Cavalry that was composed entirely of Marylanders. It was hoped that with his Maryland roots, Johnson would be able to recruit additional troops for the Confederates from the Frederick area. The small ill-trained, mostly inexperience 8th Army Corps of General Wallace was all that stood between Lieutenant General Early's combat hardened troops and the lightly defended Union capital of Washington.

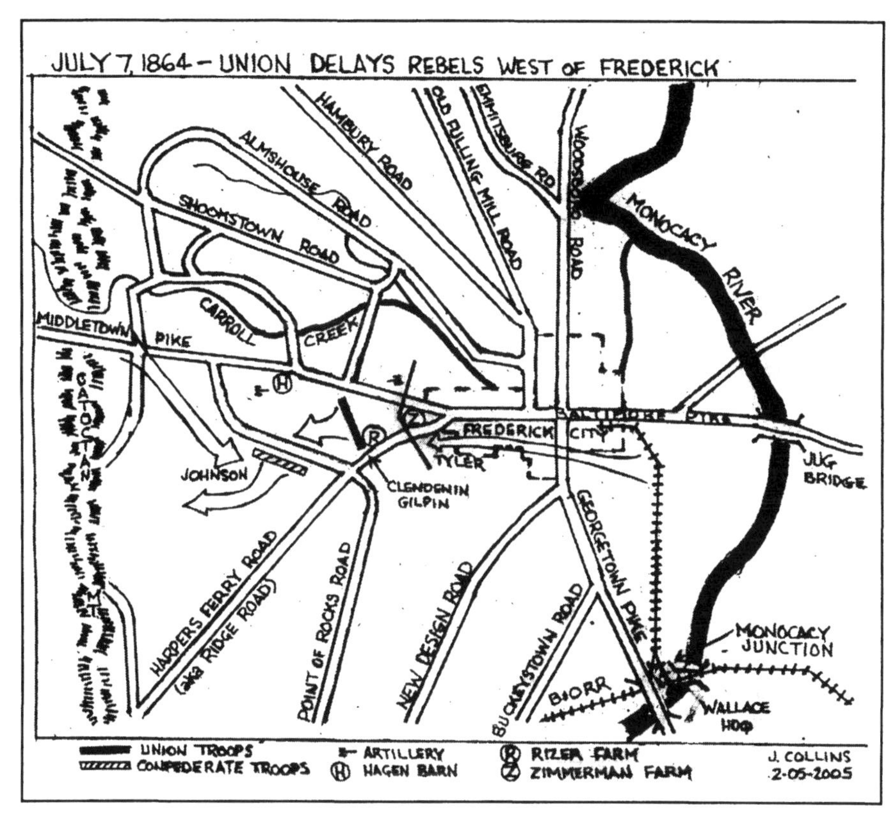

Figure 3. Map of Engagement of July 7, 1864 West of Frederick.
Map Illustrated by J. Collins

The 8th Illinois Cavalry consisting of 230 men under the command of Lieutenant Colonel David R. Clendenin was scouting the area about 5-miles west of Frederick when they encountered Major John Mosby's Ranger-Cavalry, accompanied by two artillery pieces. Clendenin reported to Brigadier General Erastus B. Tyler, who General Wallace had put in charge of the defense of Monocacy Junction, of his engagement with a superior Confederate cavalry force approaching Frederick through the Catoctin Valley. Clendenin reported he believed they were the advance guard of General Early's forces. After the brief encounter, with neither side gaining an advantage, Clendenin reported he was outnumbered and was withdrawing to a better position at Solomon's Gap located just west of Frederick. Prior to the outbreak of the Civil War, Clendenin had been a member of the "Clay Guards," a volunteer group in Washington, D.C., that protected government buildings and patrolled the city streets at night from local supporters of the Confederacy. With the start of the war, Clendenin was named an officer and helped in the formation of the 8th Illinois Cavalry.

With this information, General Tyler dispatched Colonel Gilpin and the 3rd Potomac Home Brigade, including the Van Sickles and Sergeant Davis, along with three artillery pieces of Captain Alexander's Baltimore Battery to support Colonel Clendenin's beleaguered cavalry. Recognizing the fragile Union position, the reinforcements were transported the 3-miles to Frederick by special train with specific orders to drive the Confederates from their positions back to Catoctin Mountain. Gilpin reported in his dispatch to Tyler that "have taken position on the hill west of town with enemy in full view and estimated the enemy strength at 800 men."[19] An estimate was later to be proven very low.

This was a much different situation than Company D and the 3rd Potomac Home Brigade had endured at Harpers Ferry back in September 1862. In the upcoming engagement, they faced the enemy on a fairly level playing field as they had supporting cavalry and artillery and capable leadership—things that were sadly missing at Harpers Ferry.

One can only imagine what General Bradley Johnson and his troops thought as they reached the crest of Braddock Heights. To their rear was the tranquil and beautiful Middletown Valley, while before them was the rolling, rich farmland nurtured by the Monocacy and the Town of Frederick with its skyline of clustered spires. This was the gateway to Washington, D.C. and the successful completion of their mission. One can only imagine that the weary, hungry, foot sore but battle-hardened Confederate troops for a moment thought they were gazing at their beloved Shenandoah Valley. Those pleasant thoughts of home probably quickly evaporated with the rattle of swords, clouds of dust, heat, and aching feet rudely reminded them that they were fighting for their lives.

In the early afternoon of Thursday, July 7, the opposing forces assembled west of Frederick engaged in a spirited artillery duel lasting several hours. The Confederate, four gun battery was placed on the Hagan farm about a mile west of Frederick but the advantage went to Captain Alexander's superior artillery which consisted of rifled "Parrott" guns that had a 200-yard range and were very effective in counter-battery firing. Isaac Van Sickle, his brother, cousins, and brother-in-law along with the soldiers of the 3rd Potomac Home Brigade soon found themselves involved in a hot and spirited skirmish that lasted the better part of 5-hours with the fighting centered around the wheat, cornfields, and wooded areas of the Rizer and Zimmerman farms. Although outnumbered and running low on ammunition, the Union forces repulsed three enemy charges while grudgingly falling back towards the City of Frederick. Just when the battle appeared to be over and the Union troops were beaten, Colonel Gilpin regrouped his battle weary troops and with Company D and the 3rd Potomac Home Brigade leading a final counter charge they broke through the Confederate center located at the Rizer barn and by nightfall had driven the Confederates back to the foot of the Catoctin Mountain. While casualties to the Union defenders were light, serious damage was done to the area farms, crops and livestock. As the scene of major fighting the Rizer barn was riddled with holes from shells and minie balls.

After the engagement of July 7, General Wallace telegraphed to General Halleck's command that, "...think I have had the best little battle of the war. Our men did not retreat but held their own. The enemy was repulsed three times. The force engaged on our side were the 3rd Potomac Home Brigade, two hundred and fifty men; 8th Illinois Cavalry, Lieutenant Colonel Clendenin commanding; three guns of Alexander's battery under his command, and several detachments including the one hundred days' men, Captain Lieb commanding. The fight began at 4:00 p.m. and closed at 8:00 p.m., Colonel Gilpin of the 3rd Potomac Home Brigade in direct command. From best information, the rebels were commanded by Bradley Johnson. Losses unknown. This is not official."[20]

During the skirmishing, Sergeant Davis was moving among the members of the 3rd Potomac Home Brigade encouraging them to keep up their fire. Suddenly, Davis let out a cry of shock and disbelief as he clutched his left hand. One of the multitude of minie balls whistling past the Union defenders had found its mark. Sergeant Davis had been shot in the left hand, something that Isaac witnessed. The flow of blood was quickly stopped with Davis refusing care at the Brigade hospital as he preferred staying with his relatives in the 3rd until the Confederates were driven back. Later that evening, Davis left the Van Sickles and began his trip to Annapolis being admitted to the Division One USA Hospital on July 9. Davis had the damaged finger on his left hand amputated and was returned to duty July 19; too late to join the 3rd Potomac Home Brigade in its brief adventure into the Shenandoah Valley with General Sheridan.

It should be noted that in General Tyler's written report of the engagement dated July 14, he made special mention of the 3rd Potomac Home Brigade as having conducted themselves in the most gallant manner and that their action was both brave and creditable. The soldiers of the 3rd Potomac Home Brigade, including the Van Sickles with uniforms covered with the grim of battle and soaked with sweat from the day's fighting, spent the night sleeping with their weapons at the positions they had so vigorously defended. Colonel Gilpin had fallen asleep at the nearby home of Frederick Lambert when he was awakened to give a report to General Wallace on the status of his troops and position. Gilpin reported that his troops held the favorable terrain and would hold that position as long as possible. The fighting had been so protracted that Wallace advised his superiors in Baltimore that Gilpin had reported his men being low on ammunition especially for the Sharps rifles used by the 8th Illinois. Wallace was advised that a supply of 100 rounds of ammunition per man would be dispatched by a fast passenger engine from Camden Station that evening and would be guarded by a company of infantry.

General Tyler then received word that the repulsed Confederate forces of Johnson and McCausland were being re-enforced. Be ordered three companies of the 144th and seven companies of the 149th Ohio National Guard to re-enforce the 3rd Potomac Home Brigade and 8th Illinois on Friday morning of July 8. The re-enforced Union forces continued to occupy the positions that they had won in the hard fighting on July 7. However, the action of July 8 was mainly confined to cavalry skirmishes until the 8th Illinois was ordered back to Monocacy Junction by General Wallace. As pressure from the re-enforced advancing Confederate forces increased on the tired Union defenders and when combined with the large dust clouds from three columns of advancing Confederate infantry coming through the Catoctin Mountain passes, General Wallace made the prudent decision to order the 3rd Potomac Home Brigade including Company D, the 144th Ohio, and 149th Ohio to withdraw from their hard won positions on the western outskirts of Frederick back to Monocacy Junction. Wallace had decided to consolidate his small force at this most defensible position (which blocked Early's direct approach to Washington) and await the inevitable Confederate main attack.

The 3rd Potomac Home Brigade including Company D was assigned to hold the line between the Stone Bridge and the railroad junction at an intermediary crossing of the Monocacy River known as Crums' Ford (today called Reich's Ford) located north of Monocacy Junction taking up a defensive position behind the natural embankments on the eastern bank of the Monocacy. This placed the 3rd Potomac Home Brigade between the 11th Maryland and the 149th Ohio, near the northern end of the Union defense line which stretched approximately 6-miles from Jug Bridge over the Baltimore Pike in the north to its southern terminus at the Worthington-McKinney Ford. Bolstered by the arrival of the seasoned fighters of the 1st and 2nd Brigades of the 3rd Division of the 6th Corps under the command of General James Ricketts, the Union line was defended by approximately 5,800 soldiers. To the advancing Confederates under the command of General John C. Breckenridge coming down Georgetown Pike (today's Route 355), the Union defensive line must have looked like a fragile, thin blue ribbon. The surprise that awaited the Confederates was that General Ricketts veteran troops were assigned to that part of the thin Union defensive line where the brunt of the Confederate attack would occur. Breckinridge was from Kentucky, and before joining the ranks of the Confederate Army, had served as Vice President under James Buchanan from March 4, 1857 to March 3, 1861. He lost the 1860 Presidential election as the Southern Democratic Party standard bearer to Abraham Lincoln.

With the withdrawal of the Union troops back to Monocacy Junction, Frederick was at the mercy of the Confederates. Early demanded that Frederick provide flour, sugar, coffee, salt, and bacon plus a cash ransom of $200,000 or the city would be burned to the ground. Frederick quickly capitulated with the funds from loans made by five local financial institutions (Fredericktown Savings $64,000, Central Bank $44,000, Frederick County Bank $33,000, Franklin Savings Bank $31,000, and Farmers & Mechanics Bank $28,000).[21] Early's forces left with their spoils to join Breckinridge and engage the Union forces assembled at nearby Monocacy Junction. Early's supply wagon train numbering

400 to 500 wagons, took almost 5-hours to pass through town as it followed the troops towards the battle. It is reported that the Confederate officers who stayed behind to collect the cash ransom, celebrated by eating ice cream. I think that celebration might have been premature based on the final outcome of the invasion. The citizens of Frederick estimated that the loss in materials taken by the Confederates totaled between $2 and $3 million. The City of Frederick paid off the $200,000 in loans plus approximately $400,000 in interest on October 1, 1951 (87-years later) while applying to the Federal government for repayment of the loans. It wasn't until October 1986 (122-years later) that the U.S. Congress approved money to repay the original $200,000 in loans-without interest.

The morning of Saturday July 9th promised to be another hot, hazy summer day in the Frederick area. The heat could be seen shimmering off of the weapons of both forces as they prepared for the day's events. The surrounding farm fields were golden with ripening grains and corn tassels as tall as a man's chin but soon these beautiful fields would be filled with the roar of musketry and the cries of wounded and dying soldiers.

As if to get the maximum benefit from the hot weather, the Battle of Monocacy started before 7:00 a.m. The main fighting occurred in the area of the Thomas, Baker, and Worthington farms at the Worthington-McKinney Ford of the Monocacy River, south of Monocacy Junction. The ford at this part of the river was shallow and less than 200-feet wide making it an excellent crossing point for the Confederates. The northern end of the Union lines at Jug Bridge were defended by the 144th and 149th Ohio regiments under the command of Colonel Allison L. Brown. The Ohio units engaged in brisk skirmishing with Brigadier General Phillip Cook's Georgia Brigade part of Rodes Division. Although outnumbered, they held the line against repeated attacks by the numerically superior Confederates. From their position at Crum's Ford, the 3rd Potomac Home Brigade had an excellent defensive position against a possible Confederate attack as Wallace feared the Confederates would attempt to flank his embattled Union defenders at the Junction by crossing the Monocacy at Crum's Ford. The 3rd Potomac Home Brigade had limited contact with Confederate skirmishers, who briefly tested their position and then withdrew. The flanking movement feared by General Wallace never happened. The 3rd Potomac Home Brigade had minimal involvement that was well deserved after their fighting on July 7. They were aware of the heavy fighting at the Junction as they could see the soldiers from both sides, see the smoke, and hear the cannon and musket fire.

The Confederate forces charged across Worthington-McKinney Ford with many taking their shoes off to save their shoes from being damaged by the water (shoes were a very valuable and scarce commodity) and after hours of intense fighting with the lines of both sides flowing back and forth. Finally the attackers, supported by heavy artillery fire, overwhelmed the outnumbered Union defenders of General Ricketts causing them to fallback to Georgetown Pike. Major General John B. Gordon commanding the attacking Confederate troops in his report of the engagement, wrote, " . . . the fighting was so intense that a stream running through the battlefield was red with the blood of the battle casualties for 100-yards." Recognizing the futility of sacrificing additional troops, General Wallace ordered the withdrawal of his force from Monocacy Junction. As the Union troops withdrew, they further frustrated Major Generals Framseur's Division who had fought the entire day in attempting to take the covered wooden bridge on Georgetown Pike by torching the bridge. General Wallace had ordered at 10:30 a.m. to place hay in the covered wooden bridge and burn it to further slow the Confederate march to Washington. The burning of the covered bridge must have been something to see as it was a majestic structure being 250-feet long, 50-feet wide, 16-feet high, and divided into two lanes by heavy timbers down its center. Today the Monocacy is crossed by a modern bridge where once stood the covered bridge. Lieutenant George E. Davis, Company D, 10th Vermont Volunteers led the gallant defense of both the covered and iron railroad bridges crossing the Georgetown Pike until finally withdrawing when ordered. For his actions and bravery, Davis was awarded the nations highest recognition—the Medal of Honor. During the withdrawal, the 3rd Potomac Home Brigade (including Company D and the Van Sickles) joined in helping cover the withdrawal so that General Rickett's forces (the last Union troops to leave the battlefield) could safely withdraw to fight another day. By 5:00 p.m., General Wallace's beleaguered and depleted army (now numbering approximately 4,000) including the battle hardened Company D of the 3rd Potomac Home Brigade had withdrawn from the field of battle and were eastbound on the Baltimore Pike to regroup at Ellicott's Mill. Again special note is made in General Tyler's July 14 report that the efforts of the 3rd Potomac Home Brigade on July 9 fully sustained the enviable reputation they had won in the fighting on July 7.

The following dispatch was sent by Major General Lew Wallace to the Honorable E.M. Stanton the evening of July 9, 1864, "I did as I promised. Held the bridge to the last. They overwhelmed me with numbers. My troops fought splendidly. Losses fearful. Send me cars enough to Ellicott Mills to take my retreating columns. Don't fail me."[22] Accurate troop losses from the battle were difficult to compile with best estimates placing Union losses at a minimum 1,300 while various sources estimated Confederate losses at 700 to 1,000. A very bloody battle when the total number of troops involved is considered.

In addition to the earlier detailed financial losses to the City of Frederick, the surrounding damage to farms, crops, and livestock was extensive. In the fighting at Monocacy Junction, the covered bridge over the Georgetown Pike was destroyed, as well as the nearby hotel and water tower. Two weeks after the battle, the battlefield still showed signs of the fierce conflict as evidenced by spent shells, broken cannon, and unburied corpses.

After the battle, the 3rd Potomac Home Brigade rejoined the 8th Corps at Ellicott's Mill and was then sent to Camden Station via train arriving on July 10. Isaac's younger brother David H. Van Sickle didn't stay with Company D as he was admitted to the hospital at Ellicott City, Maryland, for bronchitis and, therefore, missed the adventure with General Sheridan into the Shenandoah Valley.

Colonel Gilpin was then placed in command of Fort Worthington (one of the forts protecting Baltimore) and the 3rd Potomac Home Brigade (consisting of 581 men including Company D and the Van Sickles) along with the 1st Potomac Home Brigade, the 144th and 149th Ohio National Guard, and the 11th Maryland with forces totaling 1,461 all garrisoned at Fort Worthington. On July 14, the 3rd Potomac Home Brigade was ordered to Camden Station in Baltimore to await redeployment as part of Brigadier General John R. Kenlys' Independent Brigade and participation in the pursuit of General Early's withdrawing forces. Unfortunately, Lewis wasn't able to join the other Van Sickles in the pursuit of Early as he was again admitted to the hospital for bronchitis. He would spend the remainder of his enlistment confined to the hospital.

Company D of the 3rd Potomac Home Brigade during the Battle of the Monocacy had transformed itself from a primarily garrison unit into a cohesive fighting unit recognized for its bravery under enemy fire.

Again serving in Company D was great-great-grandfather Private Isaac Van Sickle, his younger brother Private David H. Van Sickle, three other relatives—Corporal Ephraim Van Sickle, Privates Lewis and Samuel Van Sickle, and brother-in-law John Glover, the later five having served with Company D since early in the summer of 1862, and Isaac's brother-in-law Sergeant Jefferson Davis. They had all been participants in the battle that "saved Washington" by delaying General Early's forces long enough to permit the reenforcement of Washington's defenses. Among the Union units redeployed by General Grant to defend Fort Stevens was the 139th Pennsylvania Infantry Regiment. On the roster of Company B, there was a Private named Christopher George who was a veteran of the Battle of Gettysburg. Christopher George is of special interest to me as he was also my great-great grandfather. (I hope to cover his exploits and those of his four brothers in a future book.) General Early had been surprised by the brave and stubborn defense of the Union troops that "put up one hell of a fight" at Monocacy. His troops were so exhausted from a com bination of the hot weather and intense combat that they could not begin the 40-mile march to Washington until sundown of July 10.

After a march of 2-days in the summer heat under a blistering sun, on roadways ankle deep in dust that made breathing difficult, coupled with sleepless nights caused by the same heat, General Early's forces reached the outskirts of Washington on July 12 where he found veteran Union troops (not the hoped for poorly trained reserves) manning the walls and cannon of the well constructed Fort Stevens. Early's hungry, exhausted and battle-worn force spent the next day engaged skirmishing with General Wright's Fort Stevens defenders. With an estimated one-third of Early's force capable of combat, he finally broke off contact at about 10:00 p.m. and started an orderly withdrawal of his forces from Maryland and back across the Potomac River at White's Ferry. He hoped that in the friendly confines of the Shenandoah Valley, he would be able to feed and rest his exhausted troops while planning his next move. As a sidelight, President Lincoln visited Fort Stevens during the skirmishing with minie balls reported to have whistled past his head until he was removed to safety with the help of a young army Lieutenant named Oliver Wendell Homes, Jr. Holmes was later to become a Justice of the Supreme Court.

As General Wright began to slowly follow Early's retreating forces that were slowed by the large herd of confiscated livestock and about 1,000 Union prisoners, he paused to find that the home of Post-Master General Montgomery Blair in Silver Spring and the private residence of the Governor of Maryland had been burned. On the lighter side, in Frederick, the Confederates had consumed the entire contents of Francis Preston Blair's wine cellar and took $300 in hats from N.D. Hauer's hat store.

The cost of living in Frederick skyrocketed after the Battle of Monocacy while area farmers and merchants suffered substantial losses of livestock and goods to the invading Confederates. The end of July saw coffee at 60 cents a pound, tea at $2.25 per pound, flour at $12 a barrel, potatoes $2 a bushel, wheat $2.35 a bushel, and ham at 30 cents per pound to mention a few.[23]

This was Isaac Van Sickle's second adventure in the Frederick area. As a Private in Company E of the 133rd Regiment Pennsylvania Volunteers, Isaac, while recuperating, was attached as a nurse at Camp B on Shookstown Road in early November 1862. This was a convalescent hospital and had mostly patients from the Battle of Antietam. He discharged from the army along with his brother-in-law Jefferson Davis in May 1863 having completed their 9-month enlistments.

While given only brief mention in the overall Battle of Monocacy, I believe the battle fought west of Frederick on July 7 played a significant part in helping to delay Early's attack on Washington, D.C. I hope to revisit the July 7 battle to get a better understanding of its role in thwarting the last Confederate invasion of the North.

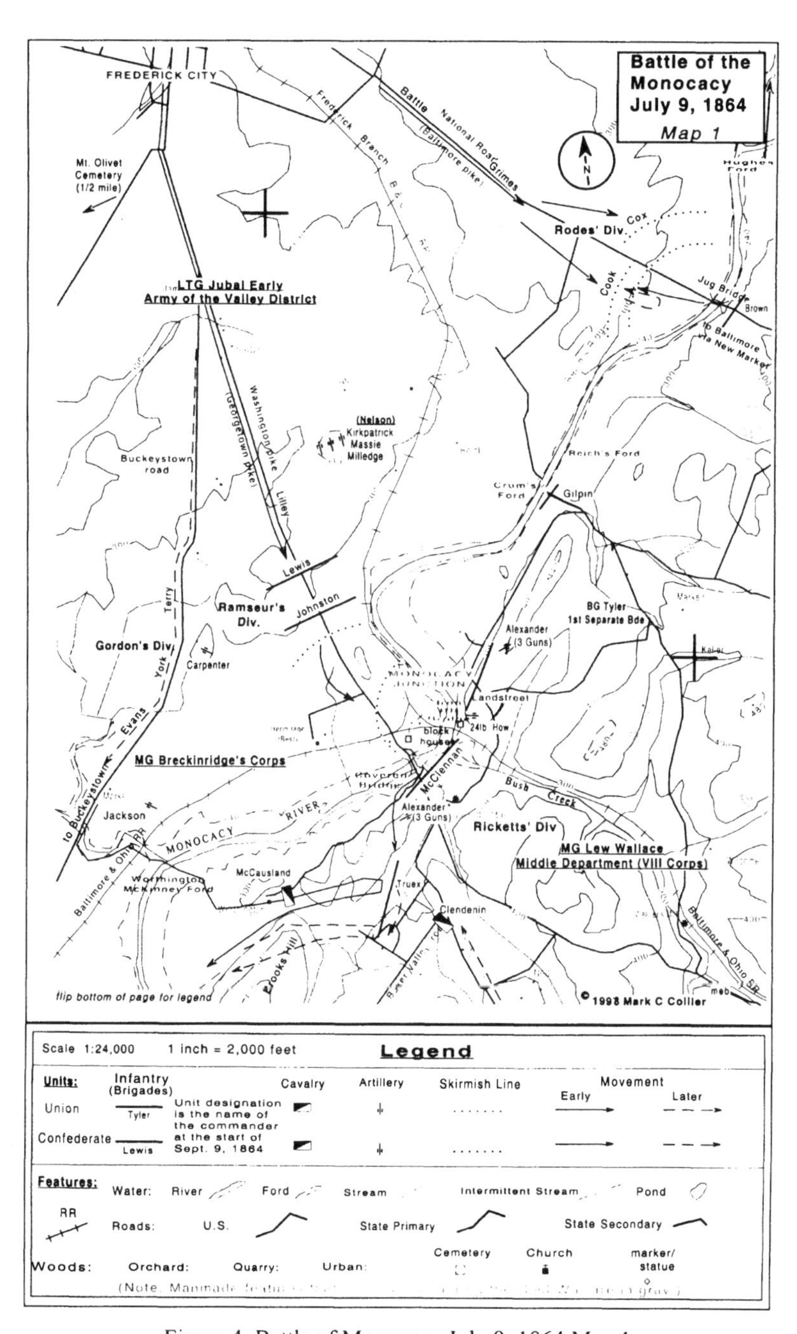

Figure 4. Battle of Monocacy July 9, 1864 Map 1

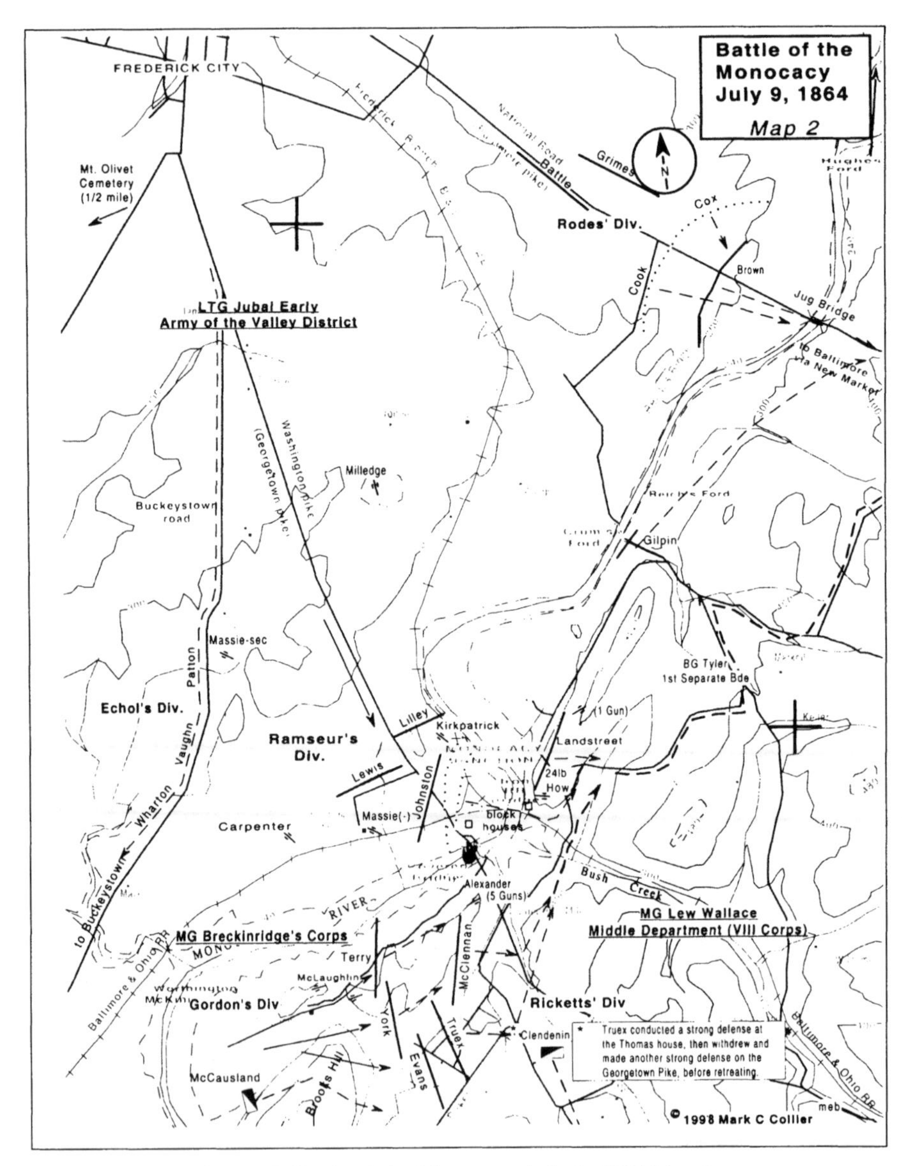

Figure 5. Battle of the Monocacy, July 9, 1864 Map 2

16. Southern Historical Society Papers, Jan-Dec 1879, Vol. IX, Broadfoot Publishing, Pages 300-301
17. Wallace, Page 718.
18. History and Biographical Record of Washington County, Maryland, Page 357.
19. Gilpin, Colonel John, B&O Telegram, July 7, 1864.
20. Wallace, Pages 732-733.
21. Gordan, Rita and Paul, A Playground of the Civil War, The Heritage Partnership, 1994 Page 173.
22. Official Military Reports of the Civil War, Operations in Northern Virginia, West Virginia, Maryland, and Pennsylvania, Chapter XLIX, Correspondence, Page 139.
23. The Diary of Jacob Engelbrecht, Historical Society of Frederick County, Inc., Page 632.

Chapter 5

After Monocacy

With Early's invasion of central Maryland and attack on Washington, the Union command recognized its continued vulnerability to invasion as long as the Shenandoah Valley was controlled by the Confederacy. Rumors had begun to circulate that General A.P. Hill's 3rd Corps was going to join with Early's 2nd Corps, and their combined forces would launch another invasion of the north. Union General Ulysses S. Grant (later the 18th President of the United States, serving two terms from March 4, 1869 to March 3, 1877) impressed on General "Black Jack" Dave Hunter, commander of the Union's 6th and 19th Corps, the importance of eliminating all Confederate forces from the Shenandoah Valley. Grant needed to concentrate all of his energies on the siege of Petersburg, Virginia, and the defeat of General Robert E. Lee's Army of Northern Virginia. This was a chance for Hunter to redeem himself from his earlier failure to stop Early's attack on Washington.

On July 14, 1864, the 3rd Potomac Home Brigade was assigned to Brigadier General John Kenly's Independent Brigade attached to the District of Harpers Ferry. Kenly's command consisted of the 2nd and 3rd Potomac Home Brigades, the 49th Regiment Pennsylvania Infantry, two cavalry units, one remount cavalry unit, and 11 artillery batteries including Captain Alexander's Baltimore battery.

On the morning of July 18, General Kenly received orders to escort a column of infantry, cavalry, and supply wagons to Winchester, Virginia, in support of General Hunter. Order of march was the 1st Division of the 19th Corps, Alexander's Baltimore battery, the 144th Ohio, the 149th Ohio, the 3rd Potomac Home Brigade, ambulances, and finally supply wagons. The 3rd Potomac Home Brigade, 144th, 149th, and Alexander's battery were all veterans of the Battle of Monocacy and eager for the opportunity to "even the score" with Early's forces.

It took the better part of the day to organize the various units into their proper order. The column finally got underway at about 4:00 p.m. on July 18. Once underway, the column proceeded without incident, arriving at a point approximately 4-miles north of Winchester, Virginia, early the morning of July 19.

General Kenly was advised that Brigadier General William Averell's cavalry brigade (attached to General Crook's 8th Corps), the advance unit of the column, had made contact with elements of General Early's forces and the 3rd Potomac Home Brigade's assistance was required to meet this threat. A brief, brisk engagement took place at Cool Springs (also called Snicker's Gap) with Captain Fallon's Company D (including the Van Sickles, minus Lewis) forcing the Confederates to withdraw. After successfully escorting the column to Winchester and participating in the action at Cool Springs, the 3rd Potomac Home Brigade and the Van Sickles returned to Harpers Ferry. Their stay there was brief as they were assigned to guard the critical Monocacy Junction iron railroad bridge. So back to the Monocacy battlefield where they stayed until early August when they were recalled to Harpers Ferry.

The Union campaign in the Shenandoah Valley then received a double blow. First, Major General Horatio Wright, Commander of the Army of the Potomac's 6th Corps, decided Early was no longer a threat and recalled the bulk of the Union troops to Washington to await further orders. Second, this reassignment of troops left Brigadier General George Crook with only three divisions and one cavalry unit to defend Winchester. When Early learned of this reduction in the Union forces at Winchester, he immediately moved to take advantage of his numerical superiority.

Confederate Generals Breckinridge and Ramseur led the attack on General Crook's outnumbered defenders at Kernstown, Virginia, just south of Winchester on July 24. After an hour of intense fighting at nearby Pritchards Hill, the Union forces broke and began a hasty, disorganized retreat north. Major Rutherford B. Hayes (later 19th President of the United States from March 4, 1877 to March 3, 1881) commanded the 23rd Ohio Brigade at Kernstown which distinguished itself during the fighting.[24] A young aide, Lieutenant William McKinley (who was elected twice as President of the United States in 1896 and again in 1900) galloped through intense fire to warn the 13th West Virginia "Mountaineers" to withdraw to safety from an exposed position. The defeat at Kernstown was very embarrassing for General Crook and the Union Army with Union losses double those of Early's forces. As Crook's beaten troops crossed the Potomac River near Williamsport, Maryland on July 26, Early continued his march north with his supplies replenished from the spoils garnered in the victory at Kernstown.

Pressing his advantage, Early, with General McCausland's cavalry ranging far in advance of the main body, continued north. He brushed aside limited Union resistance at Martinsburg, Virginia, Williamsport, and Hagerstown, Maryland, and Mercersburg, Pennsylvania while creating havoc for Union rail transportation by destroying railroad track of the B&O at multiple locations. At this point, Early keyed on the lightly defended town of Chambersburg, Pennsylvania. If he could collect a large ransom, those funds could be used to purchase desperately needed supplies and equipment. On July 30, Early's forces entered Chambersburg demanding a ransom of $500,000 in Yankee greenbacks or $100,000 in gold or they would burn the town. It was explained to the citizens of Chambersburg that these demands were being made as compensation for the destruction of Virginia property including the burning of the Virginia Military Institute by Union General Hunter. The terrified citizens could not produce the funds demanded and had to standby and watch their town set ablaze with an estimated 275 homes and 125 stores and barns being burned.

While Early was terrorizing the populace of central Maryland and south central Pennsylvania, President Lincoln and General Grant were attempting to develop a solution to this continuing embarrassment to the Union. The 6th Corps including the 3rd Potomac Home Brigade was ordered to intercept Early's marauding forces before they reached Chambersburg and drive them back across the Potomac River. The Union troops began what was to be a long march of frustration under the unrelenting summer sun. With temperatures exceeding 100 degrees, the intense heat took its toll on the Union troops. Hundreds dropped by the roadside from sunstroke with some dying. By the time the Union troops arrived at Chambersburg, the city was in ruins and Early's forces had melted into the shadows of the countryside. The 6th Army Corps then turned south and at a more leisurely pace, marched to Harpers Ferry arriving on August 2.

McCausland's cavalry continued to plague central Maryland with raids on Hancock (twice) and Cumberland before crossing the Potomac and into the relative safety of the Shenandoah Valley. McCausland's movements could only be described "as like quicksilver" as the Union's 6th Army Corps forces were only able to engage in nondecisive skirmishes with their elusive foe.

With the panic caused by Early's raids in central Maryland and south central Pennsyslvania, coupled with the inability of General Hunter to secure the Shenandoah Valley for the Union, an angry General Grant decided a change in military leadership was needed. Grant was also concerned about the overall moral of the populace in the north. Grant, not one to be slowed by governmental red tape, on August 7 named a 33-year-old cavalry officer, General Philip H. Sheridan, as Commander of the new Middle Military District. Sheridan was an Irishman who was aggressive, a careful planner, and an inspirational leader who would turn out to be the medicine that the Union would need to stop Early and secure the Shenandoah Valley. His new army totaled over 40,000 troops with the 6th and 19th Corps forming its heart. Sheridan took several days to organize his staff, units, and logistics before moving his army into the Shenandoah Valley. One of his first moves was to name old friend, General George Crook, as Commander of the Military Department of West Virginia. It is interesting to note that Generals Grant and Sheridan met at the Monocacy battlefield to discuss strategy for Sheridan's valley campaign. Sheridan decided that Winchester, Virginia, would be his headquarters. August 10 found the advance units of Sheridan's army just north of Winchester. The initial encounter between the veteran Early and new northern commander Sheridan took place on August 11 at Cedar Creek, Virginia. A young cavalry officer, Major General George Armstrong Custer, Commander of the 3rd Cavalry Division, took part in the action for the north. Since Early had established a strong defensive position at Fishers Hill, Sheridan withdrew his main force to his new headquarters in Winchester. The only problem was that Sheridan needed supplies for his army.

On the morning of August 12, General Kenly received orders to escort a desperately needed supply train from Harpers Ferry, West Virginia to Sheridan's forces encamped north of Winchester by the morning of August 13. This supply train would prove to be a monumental task, complete with many problems and hazards. To further complicate the matter, rumors placed the notorious Lieutenant Colonel John Mosby, nicknamed the "Grey Ghost," Commander of the 43rd Battalion Virginia Cavalry, in the vicinity. His Partisan Ranger cavalry had a reputation as a fast hitting, opportunistic group that did not follow conventional rules of warfare. As the wagon train full of supplies entered that part of the valley called by the locals Mosby's Confederacy—an attack by Mosby was a definite threat.

The orders as to the organization of the supply train were very specific. Sheridan's 6th Corps wagons were to be first, second the wagons of General Crook's 19th Corps, third the wagons for the Army Department of West Virginia, fourth the cavalry corps wagons with Kenly's wagons, and General Emory's artillery battery bringing up the rear. Total number of wagons in the train was estimated at a minimum of 525 wagons.

Colonel Gilpin's 3rd Potomac Home Brigade, along with the 144th Ohio and 149th Ohio (totaling approximately 3,000 men), were to serve as escorts/guards for the wagon train. Selected to lead the wagon train were two companies of the 3rd Potomac Home Brigade with Captain Fallon's Company D and the Van Sickle boys to be the lead unit. The remaining companies of the 3rd Potomac Home Brigade were placed one company for every 20 wagons. The Ohio companies were assigned to the rear of the train with two as rear guard and remaining companies spaced one for every 20 wagons. With the huge number of wagons in the train, the infantry escort/guard provided was inadequate. This was especially true when one considers the demonstrated fighting capabilities of Early's forces and the reported presence of Mosby's cavalry in the area.

Union supply wagons were big, bulky, and slow moving contraptions that were either pulled by a team of six horses or six mules and handled by a teamster. The mules—a lead (saddle) mule, swing mule, two wheel mules and two center mules. Because of limited capacity and slow speed, wagon trains were usually employed for shorter hauling of supplies with a maximum of 50-miles. A wagon with its team measured 40-feet in length.

A wagon could carry a maximum of 2,000 pounds under ideal conditions. Each soldier in the field required 4-pounds of wagon capacity per day (ammunition, food, and medical supplies) translating into one wagon per 500 men. Then consideration must be given to the fodder required for both the animals attached to the army and the teams needed to pull the wagons. Animals needed 40-pounds of fodder per day. In summary, to supply a regiment of 1,000 men with one day's rations would require 153 wagons, 230 men, and 918 mules and/or horses. Then you factor in poor condition of the roads, weather, the animals, and humans (teamsters); wagon trains were an adventure.

The wagon train was to begin at Harpers Ferry. The train would then head west to Charletown, south to Berryville, and then west to its final destination of Winchester—a distance of approximately 24-miles and a good days travel for the bulky, fully loaded wagons. The elements that made up the train were situated in Harpers Ferry and adjacent Bolivar Heights. Bringing them together into one cohesive unit was further complicated as each unit had its own quartermaster. It was probably about 10:30 a.m. when the first wagons, escorted by Captain Fallon's Company D, 3rd Potomac Home Brigade, left Harpers Ferry and probably around 1:30 p.m. when they arrived at Halltown which was situated between Charlestown and Harpers Ferry. General Kenly was to remain with the head of the wagon train but with the organizational problems created by the different units and their quartermasters, he believed it necessary to stay behind until satisfied that the organizational problems had been resolved. Although the quartermasters worked hard to get their units organized and moving while advance units of the wagon train were passing Halltown empty wagons could still be found sitting in Bolivar Heights.

It is estimated that the majority of the wagons were on the road by 4:00 p.m. and General Kenly had left the rear to rejoin the head of the wagon train. Captain Russell, quartermaster for the 6th Corps had agreed to stay with the rear of the wagon train. Once underway, the mammoth wagon train of al least 525 wagons was probably 6-miles in length and took 2- and one-half-hours to pass a given point.

The wagon train rumbled through Charlestown and Halltown, West Virginia, and proceeded south on the Berryville Pike towards Berryville, Virginia, and then onto Winchester. The wagon train stopped twice on its way to Berryville in order to refresh the men and permit the stretched out wagon train to close-up. At about 10:00 p.m., Captain Fallon reported that his men were tired and needed a brief rest but were ordered to continue to move forward. As the wagon train continued, General Kenly realized his error in judgment and a short rest was needed or the wagon train might not make it to Winchester at the time ordered. At 11:00 p.m., he ordered the wagon train to halt, giving his fatigued men a short rest. Captain Fallon put out pickets with the remaining troops taking a much needed breather; some too tired to even make coffee.

After a 30-minute break, the wagon train resumed its march only to be slowed several miles north of Berryville, Virginia, by a large tree that had fallen across the road. This forced the wagon train off the road in order to go around the obstruction. As darkness descended, the pace of the train became agonizingly slow. Problems continued as the wagon train next came to a damaged bridge crossing Long Marsh Run that would not support the weight of the heavily laden wagons. The 20 man pioneer scout unit was put to work repairing the bridge while a temporary alternate crossing was sought. A ford in the narrow stream was quickly found and Captain Fallon ordered some of his men (including Isaac Van Sickle and his relatives) to deploy as pickets on the other side of the stream. This was the second blockage of the road and the men wondered if they were accidents or deliberate. Could a force of Confederates be concealed in the darkness along the road waiting for just the right time to attack the Union wagon train? To the relief of Isaac and the others, the only thing that awaited them on the other side of the stream was empty woods.

The bridge was repaired in less than 1-hour which was good news. Only about 20 wagons had crossed the stream before a wagon broke down in the middle of the ford halting the movement of the remaining wagons. Again the wagon train was underway but it was again slowed, this time by a solitary sutler's wagon (a wagon used by an army follower selling provisions, liquor, and other goods to the soldiers) on the road. The wagon was being plundered by several Confederate stragglers. Captain Fallon's Company D quickly reprimanded the stragglers and moved the wagon to the side of the road so the wagon train could continue.

Captain Fallon's troops, including the Van Sickles with the lead wagons, passed through Berryville and continued west towards Winchester crossing the Opequon Creek at sunrise on August 13. General Kenly watched the first 50 wagons cross before riding forward 3-miles to the top of a nearby hill to get a view of Winchester. He was shocked not to see any signs of Sheridan's encamped army. In the meantime, wagons continued to arrive and began to stack up behind the stopped wagon train. At that point Colonel Edwards, commanding a brigade of the Union 6th Corps, reported to General Kenly that General Sheridan had moved his troops into the field and that Colonel Edwards was to take charge of the train upon its arrival. Edwards directed the wagons to proceed into Winchester with instructions that the animals were to be fed and watered. One of General Kenly's aids reported that the wagons were arriving in an orderly fashion although there were difficulties with the rear of the wagon train. At this point the road-weary men of Captain Fallon's Company D and the rest of the 3rd Potomac Home Brigade were ordered into Winchester as garrison troops while Colonel Edward's troops handled the arriving wagons.

Unbelievably as wagons were arriving at Winchester, some cavalry forage wagons had still not started the trip. Many were being loaded while others were still empty as the previous months of military campaigns in central Maryland had left the countryside short of forage.

Lieutenant Colonel Miller was commanding the rear guard for the wagon train and sent a report that the cavalry wagons had stopped north of Berryville at Buck Marsh Baptist Church. The animals had been unhooked from the wagons and were feeding. This stoppage had created a gap in the integrity of the wagon train and reduced the already limited firepower for the defense of the wagon train. If reports about Mosby's Rangers were accurate, the stopped wagons were sitting targets for any Confederate troops in the area. To further complicate matters, many of the teamsters for the wagons, exhausted from a long day in the hot summer sun handling the reins for a team of strong animals, had gone to sleep. One could say that this was a disaster waiting to happen. General Kenly sent word to get the wagon train moving immediately, and although Colonel Miller promptly took action to get the wagon train moving, it was too late.

As the sun began to rise, Colonel Mosby's single artillery piece opened fire on the stopped wagons. The morning was chaotic as some of the teamsters mounted their lead (saddle) mules and rode away from the wagon train. With Mosby's single artillery piece controlling the road, those teamsters who hooked up their teams found they were unable to move. Mosby's cavalry dressed in blue uniforms, armed with carbines, and led by a man in civilian clothes swooped down on the unmoving wagons. Although surprised, the 149th Ohio under the command of Colonel Brown mounted a good defense. After a brief and deadly skirmish, Mosby's Rangers retired with the spoils of the attack, 600 horses and mules, 200 beef-cattle, valuable stores, and over 200 prisoners. Mosby reported that he captured and/or destroyed 75 wagons while the Union reported 25 wagons damaged and/or destroyed. All wagons lost belonged to the cavalry's 3rd Potomac Home Brigade. Union losses were stated as six dead and nine wounded plus those captured while Confederate losses were two dead.

General Kenly, fearing more attacks on the train, ordered that the 3rd Potomac Home Brigade take up positions at the Opequon crossing to protect it until the remaining wagons crossed. Spare teams were sent to bring up those wagons that had lost their teams and an ambulance was sent to bring up the wounded. It took an additional 2-hours for the remaining wagons to safely cross the Opequon. By late August 13, the train was in Winchester. That is except for those wagons still being loaded with forage. They were still at Harpers Ferry on August 14.

On August 15, Sheridan returned to Winchester to re-supply his troops and review his tactical plans. He then began a fencing campaign between Early and the cautious Sheridan for the remainder of August. They began to set the scene to determine who would control the Shenandoah Valley—the breadbasket of the Army of Northern Virginia. They skirmished at Winchester, Berryville, and Opequon Creek with neither willing to commit their forces in an all out battle. On August 16, Early's troops suffered a defeat at Cederville (on flat terrain in front of Front Royal) with Brigadier General George Custer's cavalry brigade routing the opposing Confederate cavalry by leading a charge with sabers that were very effective in close combat. The charge of his brigade was described as a "handsome saber charge" by onlookers. On August 21, Sheridan took up a solid defensive position at Halltown (about 4-miles south of Harpers Ferry) and dared Early to attack. Early didn't take the bait. On September 1, Sheridan decided to break the stalemate and move on to Winchester.

On September 3, elements of both forces engaged at Berryville with Early having the initial advantage but was forced to withdraw after General Crook was re-enforced with troops of the Union's 19th Army Corps. Confederate General Lee ordered General R.H. Anderson's Division to Petersburg, Virginia, but Sheridan quickly blocked Anderson's path to Lee. He was forced to stay with Early. Once again Sheridan became cautious and another period of relative inactivity settled over the Shenandoah Valley. This inactivity didn't set well with either President Lincoln or General Grant. They ordered Sheridan to stop being cautious and use his superior numbers to clear Early from the Shenandoah Valley without delay! At the same time, Early's forces received a blow as Anderson's Division was finally able to leave the Valley to join General Lee at Petersburg. This left General Early with forces numbering less than 15,000 men.

While Sheridan was again planning his attack, the wily veteran Early moved on to Martinsburg, West Virginia, destroying the B&O track and rolling stock defeating a unit of Sheridan's cavalry on September 17 and 18. But this limited success of Early was to be short lived.

On September 19, the two armies met for the third and final battle of Winchester at nearby Opequon Creek. Early gained the initial advantage since General Wright had brought up the 6th Army Corps supply wagons. This created slow, heavy traffic resulting in delaying the arrival of General Emory's 19th Army Corps that was coming to re-enforce Wright's troops. Fierce fighting continued with Early's troops putting up a gallant fight although outnumbered three to one. With his troops in danger of being flanked, Early withdrew to a strong defensive position at Fisher's Hill about 20-miles south of Winchester. Early lost a quarter of his troops plus two of his most experienced commanders—General Robert Rodes and Lieutenant Colonel Alexander Pendleton. Union losses were over 5,000 but numbers was something that Sheridan had a lot to give. He could win a campaign of attrition with Early. Again General Custer and the 2nd brigade of the 3rd Calvary Division bested Lomax's Confederate cavalry utilizing the saber charge. At this point in the war, the Union cavalry was able to defeat the Confederate cavalry consistently as in close fighting—the saber was superior to carbine with pistol. Custer's habit for attacking without plans resulted in his cavalry units having the highest losses to all similar units while he, personally, had 11 horses shot out from under him during the war.

Sheridan's troops pressed on to Fisher's Hill and, with superior numbers, were able to overwhelm Early's fatigued troops on September 21 but only after another spirited battle. This battle is of particular significance as Early was close to victory in the early fighting. However, Sheridan hearing of the precarious situation of his forces, galloped 14-miles back from Winchester to the front taking control of his forces—quickly turning a potential defeat into a resounding victory. This pointed out the best military quality of Sheridan—he made the best of every action. What could have been a humiliating defeat for the Union, turned into one of the more decisive Union victories of the war. Although Early's troops were in total confusion and thoroughly routed the failure of General Averell's cavalry to follow up, permitted the retreating Confederates, now numbering less than 9,000, to withdraw into the friendly confines of the Blue Ridge Mountains through Rockfish Gap where they would continue to harass the Union troops for the remainder of the war.

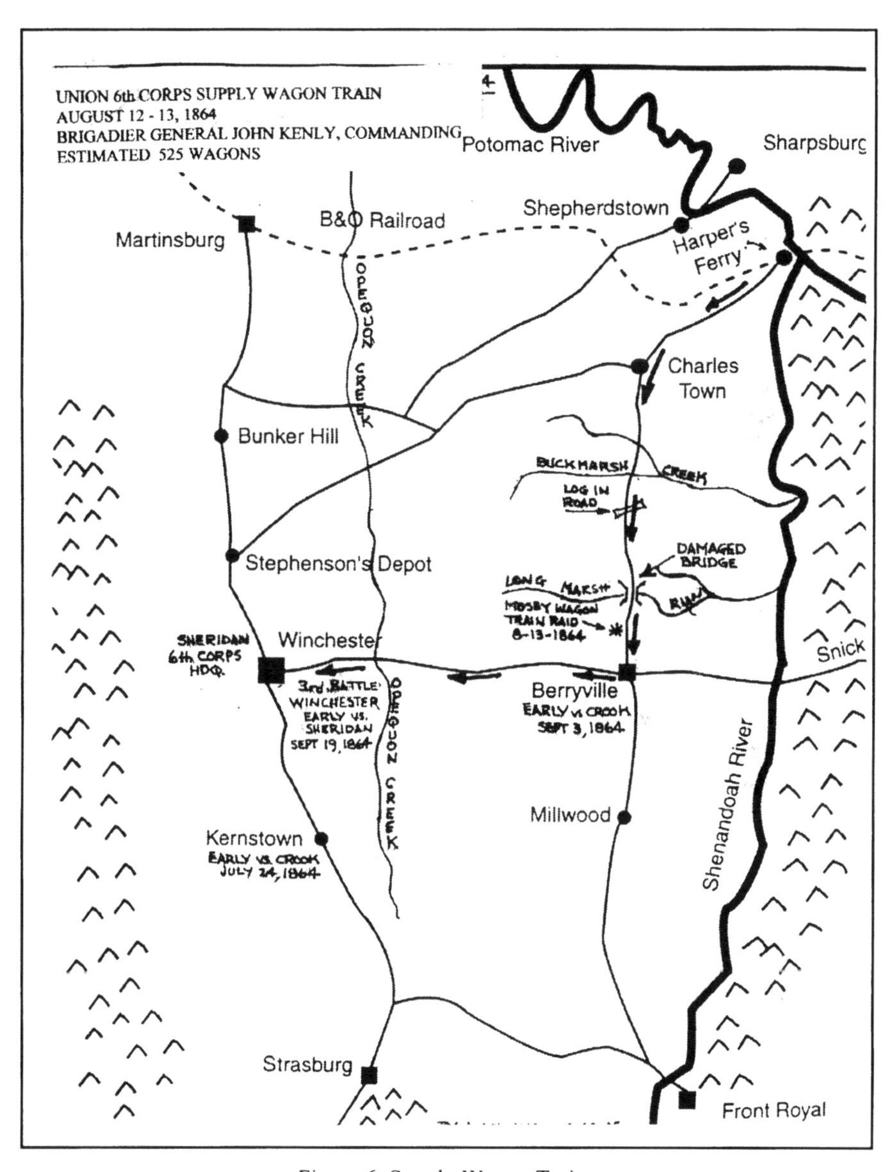

Figure 6. Supply Wagon Train.

After these two engagements, Early could no longer challenge Sheridan in a head-to-head battle as he had lost over 20 artillery pieces, almost 5,000 men, and two of his most experienced commanders. Going forward, he could only harass Sheridan. To prevent any further possible invasion by the Confederates through the Shenandoah Valley, Sheridan instituted a scorched earth policy. He burned over 70 mills filled with flour; over 2,000 barns many filled with wheat, hay, and farming equipment; stores and farm fields; plus taking over 4,000 head of stock. From Winchester to Staunton, a distance of 9-miles, the once beautiful and bountiful Shenandoah Valley and the adjoining smaller Laurel and Little Fort Valleys were filled with fires and blanketed in smoke. The pall of disaster hung over the Valley and the month of October became known as "Red October" because of the glow in the sky from the burning fires. In fact, Sheridan was heard to say about his devastation to the Valley that, "a crow could not fly over it without carrying his rations with him."[25] The agricultural breadbasket of Lee's Army of Northern Virginia no longer existed.

General Early had managed to fight 72 engagements and travel over 1,700-miles during a period of only 5-months in 1864 with that march north beginning in May after he and his troops had fought in the bloody Battle of the Wilderness. He had invaded central Maryland, ransomed and burned towns, disrupted vital rail traffic, ravaged the C&O Canal, and marched to the gates of Washington and, for a short time, diverted General Grant's attention from the siege of Lee at Petersburg. After his defeat, Early was relieved of his command by an unforgiving Confederacy.

The army wasn't going to allow Mosby's attack on the wagon train of August 12 to go without an official Board of Inquiry. Headed by Brigadier-General John D. Stevenson, the Board was to investigate the problems encountered in the movement of the wagon train, its subsequent attack, and loss of life, animals, wagons, and equipment. After meeting for 2-months, interviewing dozens of personnel, and reviewing a mountain of written reports, the Board report, dated November 13, 1864, was unable to find fault with General Kenly or any other of the officers involved. The Board found that the guard for the train was insufficient for the number of wagons in the train and that not enough pickets were posted to guard the train when it was halted.[26]

What became of Captain Fallon, Company D, the Van Sickles, and the 3rd Potomac Home Brigade? The 3rd Potomac Home Brigade was released by General Sheridan after their escort of the wagon train to Winchester and were returned to Harpers Ferry to become part of the West Virginia Reserve Division under the command of General Kenly. In September 1864, the Van Sickles and the 3rd Potomac Home Brigade began a series of marches that totaled 80-miles, ending at Clarksburg, West Virginia. There they boarded a train that took them into the West Virginia mountains to their final assignment of the war—Buckhannon where they finished their army service guarding bridges, railroad tracks, and military buildings without further exposure to the horrors of battle.

Ephraim had been battling serious diarrhea since his parole at Annapolis and was finally hospitalized for a month in October 1864. He was released from the hospital on November 3 and given a 15-day furlough so he didn't return to duty after the furlough he was classified as a deserter. However, when he did return to the 3rd Potomac Home Brigade early in December 1864, he was returned to duty without any disciplinary action. This was probably the result of his faithful service of more than 3-years, the large number of family members in the 3rd Potomac Home Brigade, and the fact that the trip through the mountains of West Virginia was very demanding for a soldier who was still recovering from the debilitating effects of diarrhea.

On May 12, 1865, the 3rd Potomac Home Brigade was shipped to Camp Carroll in Baltimore to await musturing out which finally happened on May 29, 1865. Even at this point, the Van Sickles still had to deal with the military. The muster rolls showed that Isaac owed $1.12 for lost ordinance stones while brother David owed $1.12 for lost ordinance stones and 90 cents for a lost haversack and canteen. Lewis owed $1.45 for ordinance stones and 41 cents for a canteen while Samuel owed $1.12 for ordinance stones and 90 cents for lost haversack and canteen. The same muster rolls indicated that they hadn't been paid since August 1864 and Isaac was still owed $200 of his $300 enlistment bonus.

After 3-years of meritorious service, the Van Sickles returned to their families and civilian life. They all carried the scars of the Civil War and would spend many future years dealing with the growing bureaucratic red tape of the government in securing the meager monthly pension payments that they deserved for the ills they developed while in service of their country. This delay in payment wasn't an isolated incident. Although, soldiers were to be paid monthly, it wasn't unusual for soldiers to go unpaid for as long as a year. Constant movement of units, shoddy recordkeeping, difficulty in transporting large amounts of cash and internal problems in the quartermaster corps were some of the reasons for late payments. Isaac took home with him ongoing diarrhea, rheumatism, and constant hip and back pain. Lewis, meanwhile although mustered out in May 1865, wasn't discharged from the hospital until September 1865.

For Corporal Ephraim Van Sickle, there was to be no future. His illness continued and he was again admitted to the hospital this time in Thurmont, Maryland (just a few miles north of Frederick) suffering from chronic diarrhea. Although he put up a gallant fight with his body thoroughly wasted from the effects of uncontrolled diarrhea, Ephraim died on July 10, 1865, less than 2-months after mustering out of the 3rd Potomac Home Brigade. Ephraim was another casualty of the war. More men died in the Civil War from disease and illness than from combat.

There was another Van Sickle who saw service during the Civil War. The northern counties of Virginia (later to become West Virginia) also raised troops to support the Union cause. In July 1861, the 3rd Regiment, West Virginia Infantry was formed at Clarksburg. The 3rd contained 11 companies and David

Van Sickle, the son of John Van Sickle (who was Isaac's half-brother), joined Company H of the 3rd. Unfortunately for David, his tenure in the military wasn't as stellar as that of the other Van Sickles. While Major Swearingen of the 3rd claimed that David wasn't a soldier, his poor health was probably the real reason for Major Swearingen's appraisal of David's commitment to the military as he spent more time in military hospitals than he did on active duty.

David spent the summer months of 1862 recuperating from illness in Emory General Hospital in Washington, D.C. He returned to duty with the 3rd guarding the B&O Railroad tracks, bridges, and rolling stock but after only a few months, he went absent without leave for January and February 1863. When David returned to duty after his absence, he spent the next 2-months in the Grafton, West Virginia, hospital. No sooner would David leave the hospital, he would return back to the hospital as it seemed he couldn't achieve good health. Finally on May 18, 1863, David could not continue and this single young man at 23-years of age, died from consumption. Consumption is defined as a disease that causes the body or part of the body to waste away especially with tuberculosis of the lungs. His father, John Van Sickle, received a letter from the military informing him that David was buried wearing a white shirt and drawers. David's effects were sent to his father on May 25, 1863, consisting of a blanket, dress cost, great coat, shirt, blouse, one pair of pants, one pair of boots, and $10 in Treasury notes. A sad epitaph for a young soldier and one of the Van Sickle clan.

A half cousin of Isaac named Elias Van Sickle, also served in the Civil War. He enlisted on July 10, 1861, at Clarksburg, Virginia (now West Virginia) in Company H, 3rd Regiment Virginia Infantry at age 23. His unit participated in the Battle of Cedar Mountain on August 9, 1862, and Elias was captured. Although a prisoner with the confusion that often followed battles, the company muster roll of October 12, 1862, reported him as AWOL. When Elias rejoined his unit on November 22, 1862, the confusion about his status was corrected and he was returned to duty without any disciplinary action taken. On February 27, 1864, Elias re-mustered in the service at Martinsburg, West Virginia, with a bounty of $400 due. He was promoted to Corporal on October 15, 1864, but was quickly reduced to the rank of Private after going AWOL for the month of November. He was then posted to duty at Fort Casper, Dakota Territory to help in the building of the western fort. During 1865 while working on the fort's walls, he fell and was injured. Finally, in poor health, he was discharged from the army on May 23, 1866, at Fort Leavenworth, Kansas. He returned home to spend the majority of time in bed as he suffered from rheumatism of the ankles, knees, and legs plus heart problems. Elias claimed his disability was the result of his fall and the extreme cold endured while posted at Fort Casper. Elias was living somewhere between Selbysport and Glade Farms, Maryland, receiving a pension of $30 per month when he died on April 9, 1912, at age 74.

After the war, Lieutenant Colonel Charles Gilpin relocated his family (wife Julia and four children) and business from Flintstone, Maryland to nearby Cumberland. The 1880 Cumberland census finds Charles retired and living with his wife and four children (Mary, Sophia, George, and Charles). It appears that Charles died sometime around 1893 as his wife Julia applied for his Civil War pension in that year. The same census found that Major, who had been promoted from Captain at re-enlistment, Michael Fallon had also moved to Cumberland with his young wife and child with little of his life known after that date.

24. http://www.ehistory.com/world/battleviewBID152.
25. Valley Aflame, Bryd Newspapers April 15, 2004.
26. Official Military Reports of the Civil War, Operations Northern Virginia, West Virginia, Maryland , and Pennsylvania Chapter LV, Page 632.

Chapter 6

Isaac After The War

If you remember before the outbreak of the Civil War, the 1860 census placed Isaac, Louisa, and their two small children (Jefferson and Susan) in Silbysport, Maryland. That same census also has Isaac's father David II, mother Catherine, and younger brother David Harrison in Silbysport. Occupations of both Isaac and David were farmers. With Silbysport within a few miles of Pennsylvania, it isn't hard to see how Isaac could enlist in military units from each state. In fact, during the war, many men enlisted in states that were where they didn't live because of enlistment bonuses.

As a farmer, Isaac was accustomed to the constant demands of farm life by working long hours outdoors in all kinds of weather, manhandling heavy farm implements, dealing with strong animals, and lifting heavy, cumbersome loads. When Isaac enlisted in the 133rd Pennsylvania Regiment at 27-years of age, he was a family man and in excellent physical condition.

When Isaac returned to his home and family in June 1865, he was emotionally and physically exhausted. He would forever carry with him the physical and mental scars inflicted upon him by the rigors and horrors of the Civil War. He returned to farming but found that his ability to perform hard work for long hours that farming demanded had been severely compromised by his chronic back and hip problems developed while in the army. However, he was successful in another area of his life—he fathered four more children—William Ellsworth (1862), Henry Harrison (1864), Lucinda (1866), and Missouri (1869).

The 1870 census finds Isaac, Louisa, and their six children living in Selbysport with Isaac farming and listing personal property valued at $594. During 1870, Isaac and Louisa had their seventh child—George. Then sometime in 1872, Isaac took his family north across the Mason-Dixon Line to the community of Ursina, Pennsylvania, a move of approximately 10-miles. Ursina had blossomed into a thriving community of over 1,000 spurred by the expansion of the railroad. Ursina had a variety of businesses including a grocery store, three shoe shops, three hotels, two saddlery shops, one keg (barrel) factory, and a number of boarding houses. It had two doctors, one blacksmith, and a railroad agent. The town also served the spiritual needs of its residents with three churches. It would be here that Isaac would spend the remainder of his years.

The 1870 census also found Lewis Van Sickle, a veteran of the 3rd Potomac Home Brigade, his wife Lucinda Schroyer, and six children—Samual, George, Martha, Alvidda, Andrew, and Mary ranging in age from 4- to 18-years also living in Selbysport although they would later move to Friendsville, Maryland.

Lewis Van Sickle and his wife, Lucinda Schroyer, were married by Henry Friend on July 6, 1877. The marriage occurred 13-years after the beginning of the Monocacy battle. Lucinda was the second wife of Lewis as his first wife Hulda Wulf died on March 3, 1873. Lewis had also come home from the war in poor health having spent time in three different hospitals for dysentery during his military service. While living in Friendsville, Lewis would finally lose his long battle with illness and passed away on July 16, 1888, from gangrene. His final resting place is the cemetery at the nearby Blooming Rose church.

In 1880, Isaac, Louisa, and their six children—William, Henry, Lucinda, Missouri, George, and Nora (born in 1876) were still in Ursina. As a point of interest, Isaac's daughter was named after Jefferson Davis' wife Missouri. Their two oldest children—Jefferson and Susan had left home to go on their own. great-grandmother Susan Van Sickle had married great-grandfather William Dorsey Collins and with their three young children also lived in Ursina. (That is another chapter in the Collins family history and will be covered in another installment.) Unfortunately, Isaac's ailments had progressed to the point that he could no longer endure the demand of farming. His two sons, William and Henry, were employed in the local heading mill that processed the local farmer's grain crops. Meanwhile, Isaac had been relegated to working as a night watchman for Edward Alcott.

Isaac's younger brother, David, married Sarah M. Heinbaugh of Winding Ridge, Maryland on August 19, 1866. Jefferson Davis (Isaac's brother-in-law) was a witness to the marriage. David Harrison and Sarah had six children—Jannie, George, Catherine, Mary, John, and Dorsey (ranging in age from 1 to 13) and were listed in the 1880 census of Ursina. David became one of the 23 charter members of the Ross Rush Post Number 361 that was organized July 23, 1883 in Ursina. Post Number 361 was named in honor of local son, Ross Rush of Company H, 85th Pennsylvania Volunteers who was killed on June 18, 1864 in Petersburg, Virginia.

The exposure to war, death, horrible living conditions, and exposure to germs and bacteria had a significant impact on all who served. For Lydia Van Sickle Glover, it meant that husband Harry Glover was not happy at home. Although Harry had two more children with Lydia after his return from the war, he left Lydia for another woman—Louisa Uphold (who was the wife of James Uphold). Both Harry and Louisa left their families and the area completely. The 1870 Grant County, West Virginia, census finds Lydia living with her five children (Annabelle 13, Joshua 11, Matti 7, Dorea 5, and Dollie 3). She listed herself as a widow. This was common practice of women whose husbands had left them as it enabled them to save their virtue and reputation. Lydia died from heart disease and is buried in Maplewood Cemetery. It is interesting that James Uphold was given custody of his four children after Louisa left him for Harry.

Isaac's half brother, John Van Sickle, remained in Preston County with his first wife Rachel Van Sickle who was the daughter of Zachariah Van Sickle and Susanna Markley. They had three children—Elias (born in 1836), David (born in 1839), and Susan (born in 1842). Rachel died in April 1844. John, needing a mother for his young children, married Mary Thomas Hitchens from England. They had one child Charles (born in 1849). Mary was also to pass away. John came to Blooming Rose, Maryland, and married for a third time, making Elizabeth Geary, his wife on November 28, 1866. During this time, John continued the family tradition of farming. Upon moving into Maryland, John Van Sickle purchased 228-1/4 acres of land and cultivated about 70- to 75-acres annually. The farm was aptly called the Sickle Hill Farm.

After the war, Jefferson Davis came back to Addison. The 1870 census for Addison Township listed Jefferson, his wife Missouri Crockett (they were married in 1857), and five children—James (age 12), Kate (age 10), Anna B. (age 6), John Jefferson (age 4), and Harry (age 2). On August 29, 1874, Missouri died leaving Jefferson with five young children. Jefferson didn't stay a widower long as on April 11, 1875, he married Jane Swope Lewis herself a widow. After all the moving around during the war, Jefferson let the wonder lust bug get the best of him and took his new wife and family, left Somerset, Pennsylvania, and made the arduous trip to Santa Fe in the New Mexico Territory. He would spend the rest of his years there. Jefferson was described as between 5'6" to 5'9" tall, brown hair, brown eyes, and dark complexion. He died on January 17, 1909 (73-years old) with his wife Jane following him in death on November 26, 1923.

With Isaac's health not improving, he applied to the United States Government for a Civil War military pension. He had to fill out a mountain of paperwork with the application including a statement that he suffered from chronic back and hip problems, diarrhea, and rheumatism that had been contracted while in the army. He indicated that treatment for his back problems consisted of plaster applications while a patient/nurse during his stay at Camp B in Frederick. Isaac got his brother-in-law and fellow Civil War veteran, Jefferson Davis, who was living in Santa Fe, New Mexico, to attest to Isaac's ailments. He also got a statement from his employer, Edward Alcott, that Isaac was incapable of any heavy labor or lifting, and had been relegated to the light duty of a night watchman. Under the Invalid Pension Act of July 14, 1890, Isaac was granted a monthly pension of $12. Ethel Scholl, a granddaughter of Isaac, indicated that his pension was initially paid in gold coin.[27] At that time, the U.S. had four gold coins in circulation—a $20 "double eagle," a $10 "eagle," a $5 "half-eagle," and a $2.50 "quarter eagle." They were legal tender until the discontinuance of all gold coinage in 1933.

Maintaining the pension was an ongoing process that required the pensioners to continually reapply for their meager benefits. It seemed like every year or two, the act would be amended, further complicating the pension process. On March 15, 1910, when Isaac (now age 75) applied for continuation of his pension, he was surprised to receive an increase in pension to $15 per month. He made his final continuation for his pension on May 23, 1912, and less than a year later on March 10, 1913, at the age of 78 he would die.

Louisa would then apply on March 28, 1913, for her widow's pension as provided for under the Invalid Pension Act. Louisa was receiving a widow's military pension of $30 per month when she died at the age of 88-years.

Isaac and his beloved wife, Louisa, are buried side by side in a small, hill-side cemetery located on a small dirt road off Jersey Hollow Road in Ursina, Pennsylvania. Their final resting places are marked by two granite head-stones with a Grand Army of the Republic marker beside Isaac's headstone indicating that he was a veteran of the Civil War.

Today, Ursina is a small quiet community of approximately 250 people of mostly German, Irish, English, and Dutch descent. Although Ursina no longer enjoys the important economic position it did in the early- and mid-1800's, it still retains the beauty of the valley created by the Little Laurel Creek.

Years after the Civil War was over, Isaac visited the Gettysburg Battlefield. According to his granddaughter, Ethel Scholl, when Isaac visited the Pennsylvania Monument honoring the sons of the Pennsylvania who fought at Gettysburg, he wept because he had missed the battle and didn't have his name on the Monument.[28] One should remember that this was a man of little education and a farmer who had left his young wife and small children, not once but twice during the Civil War, to help save the United States of America from destruction.

He witnessed the horrors of human carnage and desolation at Antietam as a nurse tending to the wounded. He participated in the lightly regarded but extremely important Battle of Monocacy and spent endless months protecting the vital B&O rail lines and bridges. He endured the hardships of battles, fatiguing marches, poor food, foul water, exposure to all types of weather, crowded camp life, and unsanitary conditions. Isaac was a young man who had a great love for family and country. Tens of thousands of enlisted men like Isaac, who are little remembered and mostly forgotten, were the back bone of the United States during this brutal, yet brave chapter in American history. Much is written about the commanders and politicians but without Isaac and his brethren, who knows what the United States would be like today.

27. Ethel Scholl letters of October 1979 to Joseph Collins.
28. Ethel Scholl letters of October 1979 to Joseph Collins.

Chapter 7

The 3rd Potomac Home Brigade

A review of company D's roster revealed that a total of 265 men enlisted in the unit during the Civil War. The breakdown of officers and enlisted men was as follows:

Table 1. Number of Officers and Enlisted Personnel.

3 Captains
4 First Lieutenants
3 Second Lieutenants
2 First Sergeants
11 Sergeants
15 Corporals
220 Privates
5 Musicians
1 Teamster

In addition to combat casualties, Company D had the following losses: six discharged for disability, two transferred to other companies, two resigned and 10 to desertion. When Company D mustered out on May 29, 1865, it had a compliment of 139 men. Of that number, 107 had served for at least 3-years with the other 32 men having enlisted in 1865. The total losses for the 3rd Potomac Home Brigade during the Civil War were nine killed in combat (one officer and eight enlisted men) and 74 from wounds and disease (one officer and 73 enlisted men) for a total of 83 casualties.

Table 2. Composition of Union Infantry Units.

Unit	Number of Troops	Commander
Company	100 men	Captain
Battalion	500 men/5 companies	Major
Regiment	1,000 men/10 companies	Colonel
Brigade	4,000 men/4 regiments	Brigadier General
Division	12,000 men/3 brigades	Major General
Corps	36,000 men/3 divisions	Major General
Army	144,000 men/4 corps	Major General

Table 3. Composition Artillery Units.

Unit	Number of Troops	Guns
Battery	100 men	6
Battalion	500 men/5 batteries	30

Table 4. Composition Cavalry Units.

Unit	Number of Troops
Company	100 men
Regiment	1,200 men/12 companies
Brigade	2,400 to 6,000 men 2 to 5 regiments
Division	4,800 to 30,000 men 2 to 5 brigades

Table 5. Company Composition.

Officers	3
Non-Commissioned Officers	15
Privates	82
Total	100

Combatants

Statistics for the number of combatants and casualties for the Civil War consist of a combination of Union records and estimates for the Confederacy. Although the Union kept fairly detailed records, the records that were reportedly compiled by the Confederacy mysteriously disappeared after the surrender at Appomatox Courthouse. Isaac and the other Van Sickles were part of the statistics and after reviewing numerous computations and compilations, I am providing the following estimate of combatants and casualties.

According to the 1860 census, there were slightly over 31.4 million persons living in the United States with approximately two-thirds (22.3 million) residing in those states and territories that would constitute the Union.

During the 4-years of the Civil War, slightly over 2.2 million men served in the Union army with 1.55 million enlisting for 3-year service terms. Of that number, 140,000 were killed in battle, 224,000 died from other causes, and 282,000 were wounded for a grand total of casualties at 646,000 or 29 percent of total combatants.

During the same 4-years, the Confederacy had approximately one million men of military age (18- to 45-years old) available for military service. Of this number, probably 800,000 were in uniform during the war. Best estimates are 74,000 were killed in battle, 126,000 died from other causes, and 129,000 were wounded resulting in total casualties of 329,000 or 41 percent of total combatants.

One can only guess that tens-of-thousands of veterans from both the Union and Confederacy were forced to deal with the long-term ill-effects of dysentery, typhoid fever, bronchitis, malaria, and other ailments contracted during their military service in addition to those who had been wounded and survived. As they aged, the impact of those ills would reduce their productivity, place hardships on their families, and add another cash outflow from the Federal Treasury through pension payments to the veterans.

The impact on the population of military age was staggering, as it can be estimated that combined casualties from both sides were approximately 32.5 percent. The 1870 census reflected the effects of the Civil War on the eleven states that comprised the Confederate States of America. Total population in the United States was 38.5 million, an increase of slightly over seven million individuals or 22.5 percent. The population of the eleven former Confederate states was 9.5 million or a meager increase of 400 thousand or 4.3 percent. Even by adjusting those numbers by adding West Virginia's population back into Virginia's numbers, the result is still an increase of slightly over 800 thousand or 9.1 percent. With the devastation to property and industry of the Confederate states, loss of its young men and effects of reconstruction, population growth in those states would lag behind the rest of the United States for decades.

Since the majority of Civil War military engagements were fought in those states that made up the Confederacy, devastation to their agricultural land, industry, town, railroads, bridges, and overall populace was catastrophic. The larger population, greater industrial capacity, ability to keep foreign nations on the sidelines, and stronger financial resources resulted in an advantage that the Confederacy couldn't overcome in a prolonged war. Isaac and the Van Sickle clan were part of that relentless enormous steamroller that was the Union army.

Appendix A

Bibliography

Beck, Brandon H., Civil War Battles 1861–1865; Winchester and Frederick County Virginia, Winchester-Frederick County Historical Society, 2002.

Bedford, Ned, Battles and Leaders of the Civil War, Gramarcy Press, 2001.

Bilby, Joseph, July 9, 1864: 14th New Jersey Infantry at the Battle of Monocacy, Historical Society of Frederick County, 1980.

Cartnell, Donald, Civil War Book of Lists, New Page Books, 2001.

Comeau, Dave, Footsteps Across the Confederacy, Willow Bend Books, 2004.

Cooling, B. Franklin, Monocacy: The Battle That Saved Washington, White Plans Publishing, 1997.

Davis, William C., Touched By Fire, National Historical Society Photographic Portrait of the Civil War, Black Dog and Leventhal Publishers, 1997.

Denny, Robert E., Civil War Years: Day-by-Day Chronicle, Gramarcy Press, 1998.

Frederick Civil War Heritage Trail, National Museum of Civil War Medicine, 2003.

Gannon, James P., Irish Rebels Confederate Tigers, History of the 6th Louisiana Volunteers, 1861 – 1865.

Gardner, Alexander, Gardner's Photographic Sketch Book of the Civil War, Dover Publications, Inc., 1959.

Garrison, Webb, Civil War Schemes and Plots, Gramarcy Press, 1997.

Gary, Keith O., Answering the Call: The Organization and Recruiting of the Potomac Home Brigade, Heritage Books, 1996.

Gordon, Paul and Rita, A Playground of the Civil War, M&B Printing, 1994.

Heidenrich, Chris, Frederick-Local and National Crossroads, Arcadia Press, 2003.

History of Pennsylvania Volunteers, Vol. VII, Broadfoot Publishing, 1994.

Hoye, Charles E., Hoye's Pioneer Families of Garrett County, Garrett County Historical Society, McCain Printing, 1988.

Jamieson, Perry D., Death in September, The Antietam Campaign, McWhiney Foundation Press, 1999.

Jones, Rev. J. William, D.D., Southern Historical Society Papers-January to December 1879, Vols. IX and XXIV, Broadfoot Publishing, 1990.

Johnsen, Michael, Encyclopedia of Native Tribes of North America, Gramarcy Press, 2001.

Judge, Josephy, Jubal Earlys' Campaign to Washington and the Battle That Saved A City, Potomac Press, 1993.

Katcher, Philip, The Complete Civil War, Cassell Military Press, 1998.

Katcher, Philip, Great Gambles of the Civil War, Castle Books, 1996.

Kenny, Hamill, Place Names of Maryland, Maryland Historical Society, 2001.

Knepper, Dr. George, The Official Ohio Lands Book, The Auditor of State, 2002.

LaPlaca, Jaclyn, Somerset County Pride Beyond the Mountains, Arcadia Publishing, 2003.

Leeke, Jim, Smoke, Sound and Fury, Strawberry Hill Press, 1998.

McKinsey, Folger and Williams, History of Frederick County Maryland Books 1 and 2, Regional Publishing, 1979.

McPherson, James M., Antietam The Battle That Changed the Course of the Civil War, Oxford Press, 2002.

Nasby, Dolly, Harper's Ferry, Arcadia Press, 2004.

Nichols, William H. 3d, Personal Narratives of the Rebellion, Vol. I, Siege and Capture of Harper's Ferry, Seventh Rhode Island Cavalry, Published by the Society, 1889.

Nofe, Albert A., A Civil War Treasury, DeCapo Press, 1995.

Nosworthy, Brent, Bloody Crucible of Courage, Carroll and Graf, 2003.

Paul, Amanda, Mount Savage, Arcadia Publishing, 2004.

Powell, Allan, Forgotten Heroes of the Maryland Frontier, Gateway Press, 2001.

Quynn, Allen G., Diary of Weather, 1857–1864.

Quynn, William R., Diary of Jacob Englebrecht (1818–1882), Frederick County Historical Society, 2001.

Reese, Timothy J., Sealed With Their Lives, Butternut and Blue, 1998.

Robertson, James, Jr., Civil War–Virginia, University Press of Virginia, 1913.

Roe, Alfred S., Monocacy: Co. A, 9th New York Heavy Artillery, Toomey Press, 1996.

Rubin, Mary H., The Chesapeake and Ohio Canal, Arcadia Publishing, 2003.

Schuldt, John W., Drums Along the Monocacy, Antietam Publishing, 1991.

Sharpe, Michael, Historical Maps of Civil War Battlefield, PRC Publishing Ltd., 2000.

Supplement to the Official Records to the Union and Confederate Armies. Vol. 26 Serial No. 38; Vol. 62, Serial No. 74, Broadfoot Publishing, 1996.

Swisher, James K., Warrior in Grey: General Robert Rode's of Lee's Army, White Maine Books, 2000.

Toomey, Daniel Carroll, Civil War in Maryland, Toomey Press, 2000.

Wallace, Lew, An Autobiography, Harper and Brothers Publishers, 1896.
Wallace, Paul H.W., Indians in Pennsylvania, No. 5, Pennsylvania Historical and Museum Commission, 1999.
War of the Rebellion Official Records, Vols. 19, 37, and 43, Broadfoot Publishing.
Williams, Judge T.J.C., and Thomas, James W., LLD, History of Allegany County Maryland, L.R. Titsworth and Company, 1923.
Worthington, Judge Glenn, Fighting For Time—The Battle of Monocacy, Burk Street Press, 1985.

Bibliography – Internet

Army Organization in the Civil War, National Park Service, retrieved January 20, 2004, http://www.nps.gov/gett/gettour/armorg.htm.
Battle of Front Royal, Heritage Enterprise, retrieved February 4, 2004, http://www.angelfire.com/va3/valleywar/battle/frontroyal.html.
Berryville-1864, eHistory, retrieved January 20, 2004, http://www/ehistory.com/World/BattleView.cfm:BID=750.
Berryville-September 3-4, 1864, American Civil War, retrieved January 17, 2005, http://americancivilwar.com/statepic/va/va118.html.
C&O Canal Aqueducts (and a Few Bridges)," retrieved March 30, 2005, http://www.iceandocoal.org/co/aqueducts.html.
Civil War Battlefield Guide—Monocacy, Maryland, Houghton Mifflin, retrieved January 20, 2004, http://college.hmco/history/readerscomp/civwar/html/cw_009001_monocacymary.htm.
Civil War in the Shenandoah Valley 1863–1865, retrieved February 4, 2004; http://www.rockingham.K12.va.us/EMS/Civil_War_in_the_Shenandoah Rockingham County Public Schools.
Conestoga Wagon, Canadian Mennonite Encyclopedia Online, Mennonite Historical Society of Canada, retrieved March 30, 2005, http://www.mhsc.ca/encyclopedia/contents/C6621ME.html
Cool Spring, Island Ford–July 17–18, 1864, American Civil War, retrieved January 17, 2005, http://americancivilwar.com/statepic/va/va114.html.
Fisher's Hill – 1864, eHistory, retrieved January 20, 2004, http://www.ehistory.com/World?BattleView.cfm?BID=153.
Fisher's Hill – September 21–22, 1864," American Civil War, Retireved January 17, 2005, http://americancivilwar.com/statepic/va/va120.html.
Folck's Mill, CWSAC Battle Summaries, retrieved February 4, 2004, http://www2.cr.nps.gov/abpp/battles/md008.htm.
Folck's Mil—Cumberland-August 1, 1864, American Civil War, retrieved January 17, 2005, http://americancivilwar.com/statepic/md/md008.html.
Fort Stevens—1864, eHistory, retrieved January 20, 2004, http://www.ehistory.com/World/BattleView.cfm?BID=781.

Fort Stevens—July 11–12, 1864, American Civil War retrieved January 17, 2005, http://americancivilwar.com/statepic/dc/dc001.html.

Friendsville Maryland, retrieved May 10, 2004, http://www.city-data.com/city/Friendsville-Maryland.html.

Harpers Ferry—1862, eHistory, retrieved January 20, 2004, http://www.ehistory.com?World/BattleView.cfm?BID=23.

Harpers Ferry NHP Stonewall Jackson, National Park Service, retrieved February 4, 2004, http://www.nps.gov/hafe/jackson.htm.

History and the Youghiogheny River, retrieved May 14, 2004, http://www.fay-west.com/youghtrail/information/history.php.

History of Bedford and Somerset Counties, Somerset County Pennsylvania Genealogy, retrieved May 13, 2004, http://www.rootsweb.com/~pasomers/hbs/chapter1.html.

History of the Canal, National Park Service, retrieved August 26, 2004, http://www.nps.gov/choh/History?CanalOperations.html.

Jubal Anderson Early, Columbia Encyclopedia, retrieved January 7, 2004, http://web8.epnet.com/citation.asp?tb.

Kernstown II—1864, eHistory, retrieved January 20, 2004, http://www.ehistory.com/World/BattleView.cfm?BID=152.

Kernstown, Second Virginia—July 24, 1864, American Civil War retrieved January 17, 2005, http://americancivilwar.com/statepic/va/va116.html.

McClellan Reacts to the "Lost Order," Timothy Reese, retrieved February 27, 2004, http://www.aotw.org/exhibit.php?exhibit_id=358.

Monocacy, Maryland—July 9, 1864, American Civil War retrieved January 17, 2005, http://americancivilwar.com/statepic/md007.html.

Moorefield Oldfields—August 7, 1864, American Civil War retrieved January 17, 2005, http://americancivilwar.com/statepic/wv/wv016.html.

133rd Infantry Flag, retrieved January 17, 2005, http://www.cpc.leg.state.pa.us/main/cpcweb/history/flags/showflag.html.

144th Ohio Short History, Daniel Masters, retrieved January 20, 2004, http://hometown.aol.com/dam1941/history.html.

The 149th New York State Volunteer Infantry, The Letters of Oliver Ormsby, retrieved September 19, 2004, http://www2.cr.nps.gov/abpp/battles/va119.htm

Opequon, CWSAC Battle Summaries, retrieved February 4, 2004, http://www2.cr.nps.gov/abpp/battles/va119.htm.

Opequon Third Winchester—September 19, 1864, American Civil War retrieved January 17, 2005, http://americancivilwar.com/statepic/va/va119.html.

Rutherford's Farm, CWSAC Battle Summaries, retrieved February 4, 2004, http://wwww.cr.nps.gov/abpp/battles/va115.htm.

Rutherford's Farm—July 20, 1864, American Civil War retrieved January 17, 2005, http://americancivilwar.com/statepic/va/va115.html.

Schuylkill County Civil War POWS, Pennsylvania Civil War Soldiers, retrieved September 19, 2004, http://www.pacivilwae.com/cwpa80pow.html.

Shepherstown - September 19–20, 1862, American Civil War, retrieved January 17, 2005, http://americancivilwar.com/statepic/wv/wv016.html.

Strategic Supply of Civil War Armies, Alan Anderson, retrieved January 20, 2004, http://www.world.std.com/~ata/stsupp.htm.

Williamsport Hagerstown—July 6–16, 1864, American Civil War, retrieved January 17, 2005, http://americancivilwar.com/statepic/md/md004.html.

Winchester III—1864, eHistory, retrieved January 20, 2004, http://www.ehistory.com/World/BattleView.cfm?BID-751.

Youghiogheny Scenic and Wild River, retrieved May 10, 2004, http://www.dnr.state.md.us/publiclands/western/youghiogheny.html.

Bibliography – Other

Antietam National Battlefied Map, Civil War Battlefield Series Trailhead Graphics, Inc., 2004.

Bond, Isaac, C.E., Map of Frederick Country Maryland, Maryland , E. Sachse and Co., Baltimore, Maryland, 1860.

Collier, Mark C., Battle of the Monocacy, Map Set July 9, 1864, Collier Map, 1998.

Collier, Mark c., Siege of Harper's Ferry: Map Set September 12–15, 1862, Collier Mapping, 1998.

Collins, Harry D., Civil War Pension File of Isaac Van Sickle and his widow Louisa (see Davis) Van Sickle, Lesson XI, Assignment No. 2.

Covey, B. Franklin, The Campaign That Could Have Changed the War– And Did, North South Magazine, Vol. 17, No. 5, August 2004.

Drenning, Dahl, Potomac Home Brigade Fought in Battle of Monocacy, July 10, 1985.

Gilpin, Colonel John, 3rd Regiment Potomac Home Brigade, B&O Railroad Telegram to General Lew Wallace, July 7, 1864.

Glades Star, The, No. 18, June 30, 1945, The Garrett County Historical Society.

Glades Star, The, No. 36, December 31, 1949, The Garrett County Historical Society.

Lake, D.J., C.E., Atlans—Frederick County Maryland, C.O. Titus, Philadelphia, PA, 1873.

Maryland State Archives, letter dated August 21, 1996 to Harry D. Collins.

Monocacy—3rd Regiment Potomac Home Brigade Infantry, Monocacy National Battlefield, National Park Service.

Muster Rolls from the 133rd Pennsylvania and 3rd Potomac Home Brigade.

Pension Files of Isaac Van Sickle and other family members.

Scholl, Ethel, letters to Joseph V. Collins.

Valley Aflame, The, April 15, 2004, Byrd Newspapers.

Van Sickle, Joseph, Letter dated March 12, 1936 to Charles H. Hoye.

Weidner, William c., Poor Ashby Is Dead, America's Civil War Magazine, Vol. 18, No. 2, May 2005.

Bibliography – Research Venues

Antietam National Battlefield.
C. Burr Artz Central Library, Frederick, Maryland.
Blooming Rose Church and Cemetery, Blooming Rose, Maryland.
Burkettsville, Maryland.
Camp Parole, St. John's College, Annapolis, Maryland.
Casseleman Bridge, Grantsville, Maryland.
Confluence, Pennsylvania.
Frederick, Maryland.
Frederick County Historical Society, Frederick, Maryland
Friend Family Association of America, Friendsville, Maryland.
Friendsville, Maryland.
Garrett Country Historical Society, Oakland, Maryland.
Harpers Ferry, West Virginia.
Monocacy Aqueduct, Chesapeake & Ohio National Historical Park.
Monocacy National Battlefield and Library, Frederick, Maryland.
Mount Olivet Cemetery, Frederick, Maryland.
National Archives, Washington, D.C.
National Museum of Civil War Medicine, Frederick, Maryland.
Oakland, Maryland.
Petersburg Turnpike Tollhouse, Addison, Pennsylvania.
Point of Rocks, Maryland.
Selbysport, Maryland.
Sharpsburg, Maryland.
Ursina Cemetery and Town, Ursina. Pennsylvania.

Index

About the Author

Joseph Collins is a retired banker and this is his initial adventure into the world of writing. He has a B.A. Degree in History from Juniata College and is currently studying for a Master's Degree at Hood College. He has always had an interest in his family ancestors, their history, and their involvement in the Civil War time period. He hopes this book is only the first installment in that history. He currently lives in Frederick, Maryland.